BETTER PRODUCTS FASTER

A Practical Guide to Knowledge-Based Systems for Manufacturers

William H. VerDuin

IRWIN

Professional Publishing
Burr Ridge, Illinois
New York, New York

IRWIN
Concerned About Our Environment

In recognition of the fact that our company is a large end-user of fragile yet replenishable resources, we at IRWIN can assure you that every effort is made to meet or exceed Environmental Protection Agency (EPA) recommendations and requirements for a "greener" workplace.

To preserve these natural assets, a number of environmental policies, both companywide and department-specific, have been implemented. From the use of 50% recycled paper in our textbooks to the printing of promotional materials with recycled stock and soy inks to our office paper recycling program, we are committed to reducing waste and replacing environmentally unsafe products with safer alternatives.

Project editor:	*Jean Lou Hess*
Interior designer:	*Larry J. Cope*
Cover designer:	*Tim Kaage*
Art coordinator:	*Mark Malloy*
Compositor:	*TCSystems, Inc.*
Typeface:	*11/13 Times Roman*
Printer:	*Book Press*

Library of Congress Cataloging-in-Publication Data

VerDuin, William H.
  Better products faster : a practical guide to knowledge-based systems for manufacturers / by William H. VerDuin.
    p. cm.
  Includes bibliographical references and index.
  ISBN 0-7863-0113-9
  1. Production engineering—Automation. 2. Expert systems (Computer science)—Industrial applications. 3. CAD/CAM systems. I. Title.
TS176.V45 1995
670'. 285—dc20   94–8032

Printed in the United States of America
1 2 3 4 5 6 7 8 9 0 BP 1 0 9 8 7 6 5 4

Preface

Better Products Faster is about knowledge-based systems, a new class of computer technologies that are revolutionizing manufacturing. These technologies provide unique and powerful capabilities that address the issues faced by manufacturers today. This book describes these technologies and the ways that they solve problems in the design and manufacture of discrete parts, as well as products made by batch and continuous processes. The focus is on problem-solving. The goal is to enable readers to identify opportunities and implement successful systems.

This book addresses that goal by explaining how knowledge-based systems solve design and manufacturing problems. First, the natures of the several distinct knowledge-based technologies are described, as are their strengths and limitations. Second, these capabilities are illustrated with applications in design and manufacturing. Examples include project benefits, costs, and paybacks. A how-to-do-it guide covers topics such as identifying, justifying, and implementing applications. A review of technology trends and a glossary round out the information presented.

Progressive manufacturers have always embraced the latest technologies. They do so to beat the competition. Computer-aided design (CAD) and computer-aided manufacturing (CAM) have become commonplace because these computer technologies provide the improvements in design and manufacturing that customers require.

Knowledge-based systems are providing significant new capabilities in design and manufacturing. On this basis, they can be viewed as the next step in CAD and CAM. Products are better designed with knowledge-based systems because they incorporate the experience of the best designers and an understanding of the best combination of geometry, materials, and manufacturing methods. Products are designed faster because knowledge from a variety of sources makes the first design better, and so time-consuming testing and redesigns are minimized.

Products are manufactured better with knowledge-based systems because the process design is integrated with the product design, providing higher quality and lower costs. The expertise and analytical capabilities in knowledge-based systems help engineers set up manufacturing processes faster. Engineers and operators benefit from expanded capabilities in process monitoring and diagnostics. Knowledge-based systems even optimize processes to improve quality, lower costs, and minimize pollution.

Knowledge-based technologies are often referred to as artificial intelligence (AI). AI conjures up many images, not all of which are positive. Some see AI as a hot research area. Some see it as weird science, a technology in search of a mission. Others see AI as an oversold technology that has not lived up to its promises.

The first and third comments are true, as they are for most new technologies. CAD, CAM, robotics—all were presented as the one true solution to all manufacturing problems. None lived up to its hype, yet all have merit. Each new technology starts with great expectations; then reality sets in. Limitations, but also strengths, are identified. Applications demonstrate the technical and commercial benefits of these strengths. In this way, each new design and manufacturing technology establishes itself and becomes accepted and widely implemented.

Knowledge-based systems are now at that point. Knowledge-based technologies have been applied to real problems in design and manufacturing. Limitations and strengths have been identified. The capabilities of expert systems, fuzzy logic, and neural networks, the three knowledge-based system technologies described in this book, have been shown to be significant and to provide capabilities not possible or practical by other means. This in turn has opened up new opportunities in design and manufacturing.

New competitors, new market requirements, and new constraints require that products be redesigned and manufactured faster and better. As manufacturers compete for market share in a worldwide marketplace, the acquisition and use of information becomes a strategic issue. Expert systems, fuzzy logic, and neural networks are in essence information-acquisition and management technologies. Thus, their information management capabilities are of strategic as well as technical significance.

This book is intended for those interested in technology and its application. This book is also intended for those whose interests lie in the changing nature of manufacturing and the role that technology plays in this transformation.

Other fine books provide detail on the research issues and technology details associated with knowledge-based systems. Much can be said in this area because it remains a hot research area with many technical improvements yet to be made. But knowledge-based systems have matured to the point where, today, they provide useful and cost-effective solutions. That is the focus of this book.

This is a how-to-do-it book. The first chapters set the stage by reviewing capabilities and limitations of current computer technologies for design and manufacturing. Knowledge-based technologies are presented from an engineering and problem-solving perspective, rather than a computer science and research perspective. Design and manufacturing applications are presented to illustrate capabilities and range of applications. The final chapter provides hints on how to identify and justify applications and how to plan and implement projects to minimize risk and maximize benefits. A glossary relates knowledge-based system terms to design and manufacturing applications.

The applications are drawn from a variety of industries and call upon a number of technologies and tools. Every effort was made in the preparation of this book to provide the widest possible coverage of applications and tools. One challenge in the area of knowledge-based systems is a direct result of its significance as a strategic opportunity. Its strategic importance results in many interesting applications remaining proprietary information and therefore being unavailable for inclusion in this or other publications. The author has included a number of applications examples with which he has first-hand knowledge. These applications are counterbalanced as much as possible with other applications to provide the broadest possible perspective.

Engineers will find this book useful as a guide to a new but relevant technology. It will help technical readers understand these technologies in a way that helps them identify and justify appropriate applications. In this way, they can participate in, and perhaps take a leadership role in, interesting and high-visibility projects.

The applications presented provide the reader valuable lessons learned in real-world implementations and help calibrate the reader's expectations of such systems. The nature of the applications presented suggest that mechanical, electrical, computer, industrial, and systems engineers as well as physicists and chemists will find the information useful for applications they may encounter in product or process development, industrial engineering, or product support.

Technical management in both design and production organizations should also find this book useful. First-level managers up through vice presidents of technology or R&D will find the material of interest on both a strategic and technical level. *Better Products Faster* will help them in their roles of identifying technology-based opportunities and defining and executing the programs to seize these opportunities. Technical managers are well positioned to recognize knowledge-based system opportunities within their organization, and so this book will help them seize the opportunities presented by those new technologies. Technical management may also choose to refer *Better Products Faster* to members of their staff, perhaps taking the common step of appointing a staff member as the in-house knowledge-based system expert charged with identifying opportunities and implementing systems.

General management may also benefit from this book. For top management responsible for corporate strategy and tactics, knowledge-based technologies are of interest as a new yet practical tool that addresses pressing problems not only in design and manufacturing but across the organization. By providing new capabilities in information management, these systems redefine opportunities to integrate organizations and in this way derive competitive advantages from the effective use of data and knowledge. Top management should understand these opportunities, as they may benefit their organization as well as the competition.

Students may also find the book useful. Those in technical disciplines will gain an understanding of application issues as a counterpart to studies of theory and technology. Students of business management, particularly those at the graduate level, may benefit by exposure to a new way of thinking about information management and the implications for the structure and operation of organizations.

Better Products Faster will expand the reader's horizons by providing the following information:

1. A no-nonsense description of knowledge-based technologies, benefits, and limitations gives readers an overview of opportunities and an understanding of the ways that knowledge-based systems relate to current technologies and problems.
2. Application examples in design and manufacturing, for both discrete part manufacturing and process industries, help readers identify new applications.
3. An engineering and problem-solving perspective, with project planning and management guidelines, helps readers implement these new and powerful technologies within their organization.

Chapter 1 begins the process with an overview of the role of knowledge-based systems in design and manufacturing.

William H. VerDuin

Acknowledgments

This book, as well as a number of the applications described in it, was made possible by the creativity and hard work of everyone at AI WARE, Incorporated, with whom I have had the pleasure to work. I thank Dr. Yoh-Han Pao and his staff for their efforts in support of this book. Beyond this, I thank them for their part in developing the technology and applications that have significantly advanced the state of the art in knowledge-based systems for design and manufacturing.

On a personal level, I thank the three people who have made my creation of this book possible: Dr. Yoh-Han Pao, Ernie Martt, and my wife Sandy.

Trademarks, Registered Trademarks, and Copyrights

1. UNIX is a registered trademark of UNIX System Laboratories, Inc.
2. NT and Windows are registered trademarks of Microsoft Corporation.
3. SPARC station is a trademark of SPARC International, Inc.
4. DEC, VAX, and VMS are trademarks of Digital Equipment Corporation.
5. Functional Link Net and CAD/Chem are trademarks of AIWARE, Inc.
6. IBM and OS/2 are registered trademarks of International Business Machines Corporation.
7. TI and Texas Instruments are registered trademarks of Texas Instruments.
8. Personal Consultant is a copyrighted name of Texas Instruments.
9. Macintosh is a registered trademark of Apple Computer, Inc.
10. HyperChem is a registered trademark of Hyperchem.
11. G2 is a registered trademark of Gensym Corporation.
12. EXPOD is a trademark of Mitsubishi Research, Inc.

Contents

Chapter One

Introduction:
The Role of Knowledge-Based Systems in Manufacturing

1.1 INTRODUCTION

Of the many technologies invented, only a small fraction are ultimately successful. Technologies may be proposed in technical papers or demonstrated on a limited scale. But only a few stand the test of time. Many technologies are not sufficiently robust to withstand the jump from small, neat demonstration problems to large, messy real-world problems. Other technologies are sufficiently powerful but are never embraced because they do not solve a perceived problem and are seen as a technology in search of a mission. The few technologies that make the cut are those that meet two requirements:

1. Provide some specific benefit over alternative approaches.
2. Address a perceived problem or opportunity.

This book is about new technologies that meet both requirements. The technologies are called **knowledge-based systems.** The several technologies under this heading are distinctly different in nature, but they share some common characteristics. They enable computers to be broader and more effective problem-solving tools; help users develop better designs and better processes by making the best design and manufacturing expertise readily available; and help users identify relationships among design geometry, materials, and production processes, and leverage these relationships to provide the best combination of product features, quality, and cost.

These capabilities address current design and manufacturing issues, and thus provide the *pull* of these new technologies into real-world applications. This *pull* complements the *push* that has been provided by several decades of research into ways to extend the capabilities of computers. The combination of push and pull has moved these technologies out of the laboratory into real applications. These technologies have demonstrated clear benefits: They help engineers design and manufacture better products faster.

These capabilities are particularly interesting because manufacturers now confront a variety of challenges. Worldwide competitors provide new benchmarks in product features, quality, and cost. Customers respond with heightened expectations. Product lifetimes shrink as more competitors enter the marketplace and as more effective design and manufacturing technologies enable more frequent redesigns and products that address increasingly narrow niches.

Environmental and other regulatory requirements provide additional challenges. Because of competition, the costs of material substitutions and process changes are often borne by manufacturers, not customers. Manufacturers also face competition from organizations less burdened by high regulatory, not to mention labor, costs.

These challenges notwithstanding, there is evidence that manufacturers can flourish in the presence of increasing customer expectations and disparities in the cost of doing business. What is required is to seize opportunities to improve both products and internal operations. This requires awareness of new needs and opportunities and the flexibility to respond in a timely and cost-effective manner. It also means making best use of available resources.

One of these resources is information. Some information comes from external sources such as new customer requirements, new regulations, new materials and technologies, and new competitive products. Information is also generated internally. Examples include analysis and plans, designs of products and processes, and manufacturing specifications and schedules. Generating these types of information requires specialized knowledge, which may be acquired either through training or experience.

Although such knowledge is important, it is often scarce. Corporate downsizing or retirements may make critical design or problem-solving experience unavailable. On the factory floor, sparse staffing and often marginal reading, mathematical, and problem-solving skills may impede troubleshooting and maintenance of the increasingly sophisticated processes called upon to produce ever-increasing levels of quality.

One response to these challenges is to rethink the role and the strategic importance of knowledge. It is increasingly viewed as a critical resource, vital to long-term success. New requirements for knowledge are emerging as a result of new products, new materials, new processes, and new competitors. Traditional sources of knowledge, such as experienced and trained staff, may not suffice to meet new needs. For this reason, new information management techniques are needed.

Organization-wide computer networks linking computers on every desk and at every manufacturing cell help provide access to information. However, as access to data increases and knowledgeable staff decreases, it is increasingly important to provide the insight required to make effective use of the data. What is needed is not data as such but the understanding of current situations, underlying causes, and appropriate actions.

Knowledge-based systems address these issues. The function of knowledge-based systems is to capture and organize knowledge and make that knowledge available for reuse in analyzing new situations and solving new problems. Thus, the role of knowledge-based systems in manufacturing is threefold:

- At a technical level, to provide a mechanism to capture, organize, and reuse knowledge.
- At an operational level, to support design and manufacturing tasks by providing ready access to expert advice and decision-support models.
- At a strategic level, to leverage in-house resources in ways that will provide clearly visible competitive advantages.

The following chapters review design and manufacturing technologies, with a focus on computer-based technologies. **Computer-aided design (CAD)** and **computer-aided manufacturing (CAM)** are important as technical reference points by which knowledge-based system capabilities can be measured. They are important as technologies that have met the two requirements for success: feasibility and need. They are presented as specific technologies and as solutions to problems. They are also important as conceptual starting points for knowledge-based systems. The following chapters on CAD and CAM set the stage for a presentation of knowledge-based systems as an emerging technology and as a proven problem-solving approach.

Chapter Two

Computer-Aided Design:
Faster Design and Easier Record Keeping

2.1 INTRODUCTION

As manufacturing technologies and customer requirements became more sophisticated, the process of design became more complex. This created the need for increasingly powerful design technologies. These needs involved the analytical aspects of design, as well as the translation of concepts and calculations into specific designs.

Since their early days, computers have supported analytical design tasks. As computers became more powerful and more available, their design applications expanded, replacing manual drafting, providing powerful analysis and simulation capabilities, and increasingly integrating design and other information throughout the organization. Knowledge-based technologies are now extending **computer-aided design (CAD)** by providing new capabilities in design and integration. Design expertise is captured and reused. Design and manufacturing problems are anticipated and corrected in the design stage. Relationships among design, materials, and manufacturing are discovered and acted upon. These capabilities may be viewed as next-generation CAD.

2.2 THE PAST

The task of design is as ancient as mankind. The first humans designed works of art that still decorate cave walls. Early product designs provided stone tools for striking, cutting, or forming. Early process designs identified eras. The Bronze Age and the

Iron Age signify the invention of processes that made new materials available. These materials made possible new products and changed lifestyles. Tools enabled humans to progress from nomads to hunter-gatherers and then to farmers. More advanced tools and processes assisted the development of villages, towns, and civilizations.

As manufactured products changed lifestyles, design played a critical role. As technologies progressed, design linked then-current capabilities to then-current needs. With the increasing sophistication of processes and products, design technologies evolved into sophisticated but paper-based engineering design and drafting techniques. These techniques addressed the separate but related design elements of creativity, analysis, and accommodation of practical requirements.

Manual drafting could be a time-consuming process requiring a steady hand and good hand-eye coordination. The drawing had to be neat and complete. Complex geometries were laboriously generated with triangles and compasses. Fine detail such as bolt threads or gear teeth, if drawn in full and accurate detail, would require hours of drafting time.

Practical requirements for a design could be complex. The design might involve complex geometries, to mate functionally and dimensionally with perhaps 20 other parts in a subassembly. If a flash of inspiration translated into a slight design change, the draftsperson started over. If next year's model was to be the same as last year's model but one inch longer, the draftsperson again had to start over.

In many cases, the development of a design required creative and analytical tasks that preceded development of the specific design. The need for different sets of skills and the often time-consuming nature of drafting led to the division of the design task into three subtasks performed by three or more individuals—an engineer, a designer, and a detailer. The engineer would direct the entire design process, defining design concepts and the physical and functional interaction between the various elements of a part or process. The engineer would also perform any calculations required to meet functional requirements.

The engineer would work with a designer, who would draw the design layout that provided an overview of the entire design, including the sizes and arrangements of, for example, all of the individual machines in a process line or all of the subassemblies in a larger assembly. Based on the dimensions and configurations shown on that layout, the designer would then design the individual machine elements or subassemblies.

The designer would then work with one or more detailers to complete the design of individual components by adding any additional views of the part required to show unambiguously all design features. The detailer would also add details and supporting documentation such as fabrication notes, specifications, and bills of material. The division of design labor depended on the available people and the size of the job, but the total manual drafting labor content justified the inefficiencies of splitting what is conceptually one task among three or more people.

In retrospect, an obvious opportunity existed to use computers to support these tasks. Computers could support up-front analysis. They could also support the conceptual task of determining spatial relationships. Fitting a part or a process element into a particular assembly or plant is accomplished by manipulating geometries: changing shapes, dimensions, and arrangements. Since many functional properties such as strength and durability are directly related to geometry, the geometry must also be revised to meet functional goals. The computer could, with proper input and output devices, replace not only manual analysis, but also manual drafting.

CAD systems were intially sold on the basis of direct labor savings. Computer-based functionality could simplify some design tasks and eliminate others, so that three separate jobs could be reduced to one or two. The engineer would still be required for the conceptual work and to identify calculation requirements. But much of the work of manipulating geometries and filling in details and dimensions could be handled so easily that fewer designers and no detailers were required.

CAD vendors claimed that productivity improvements would reduce design time to fractions of that required with manual drafting. As evidence, they showed how the designer could define, say, one gear tooth, then automatically replicate that tooth around the gear. In this particular application, the improvement in speed was impressive (assuming that the alternative is to have the designer draw all of the gear teeth rather than just to show a few).

In real life, however, CAD was not always faster than manual drafting. One example of this was provided by the General Electric Lighting Business Group. This innovative manufacturer of standard, specialty, and high-efficiency light bulbs implements advanced design and manufacturing technologies. As an early innovator, it maintains competitive advantages in a worldwide market and learns the hard way about challenges imposed by new technologies.

In the early 1980s, the Lighting Business Group installed CALMA CAD systems. These minicomputer-based systems were state of the art. Each system was built into a custom cabinet that included a large screen on which the drawing was displayed and a smaller screen displaying user commands and system response and status messages. (This smaller screen was actually a complete Lear Siegler terminal in its own cabinet inside the large CALMA cabinet.)

A keyboard, light pen, and keypad array shared space on the work surface. The keyboard was used to input system commands. The light pen was used to click on the screen to either select existing points or to define new points on the design in process. The keypad contained over 100 buttons, each with a cryptic two-letter legend. Each keypad button implemented a drafting command to implement the CALMA system's three-dimensional (3-D) CAD capability.

The 3-D approach contrasts with the less sophisticated 2-D capability, which is the direct descendant of drafting. Each part drawing consists of at least three views (typically front, side, and end views) to unambiguously describe the part. The draftsperson draws design features in each of the three views and must ensure that the features are consistent among the three views. For example, if a hole diameter is shown in a front view, the length and centerline of that hole would be seen in the top and side views. In manual drafting and in 2-D CAD, the designer must recognize the need for representation in multiple views and then implement appropriate features in each view.

A 3-D CAD system, in contrast, understands the relationship between the three views. For example, a hole drawn by the user in the front view would automatically be shown in the top and side views, based on the CAD system's built-in understanding of the geometric relationships among the three views. The user would only clarify details that the CAD system could not calculate.

The point of 3-D was to save time. The price to be paid in the early 1980s was that the system was difficult to learn. Good designers took six months to learn the system—and in particular, the use of those many, many keypad buttons—well enough to achieve their previous levels of productivity. Such was the state of the art. Ten years ago, the interfaces between people and computers were not friendly. The CALMA demonstrated an early and unsuccessful attempt to provide users both power and flexibility. It remains a challenge to provide sophisticated, powerful, computer-based capabilities in a user-friendly way. Newer systems approach this goal with user interfaces that replace many simultaneously displayed buttons with simple and nonintimidating computer screen-based interfaces that still provide access to a large number of choices.

User interfaces, and advanced technologies in general, develop only because early innovators such as GE are willing to be the first to apply a technology. In this way, the GEs of the world provide both the vital user feedback and the funding that enables technologies to mature.

2.3 THE PRESENT

Manufacturers now confront a different competitive environment, and as a result, their design needs have changed. The role of design remains vital. New designs continue to supply new products and processes that provide more features, higher levels of quality and safety, and better use of resources.

Design also serves to distinguish products. In a worldwide economy, an overabundance of products compete for the customer's attention. With many good products available in most categories, customers have the luxury of choosing products not only on the basis of objective features such as function, quality, and price, but also on the basis of subjective features such as style and color.

Meeting the changing needs of the marketplace requires effective market research coupled with effective product and process design. Designers must rapidly translate new needs into new products. As an extreme example of a rapidly changing market, it has been reported that whatever is currently hot in the Tokyo consumer electronics marketplace will be replaced by the next hot product within six weeks. Imagine the design, manufacturing, distribution, and marketing requirements to provide the response required in such a marketplace!

In more normal electronics marketplaces, change may still be rapid. Changes such as new recording media can create overnight demand for new home-entertainment products. In computers, new integrated circuit technologies lead to dramatic performance improvements but require a total redesign of the computer to fully leverage these improvements.

Automobiles also must respond to changing demands. Government regulation and market forces drive improvements in safety, durability, and emission control. Some years ago, a gas crisis consisting of temporary restrictions in fuel availability created immediate changes in customer needs. In the absence of technology to rapidly redesign and produce new products, manufacturers of large, relatively inefficient cars suffered.

Again, even in less fickle marketplaces, product lifetimes are shrinking. No longer can a design remain unchanged and successful over decades as did the Model T and VW Beetle. In more recent history, cars might be redesigned every 5 to 10 years. To do this effectively requires new approaches. The benchmark for the time required to develop a complete redesign continues to drop. Thirty months is now considered competitive, with 24 months the target. There is some question if product lifetimes will continue to shrink. Regardless, the technology that allows short design cycles enables a given staff to support more models and respond more rapidly.

Progressive manufacturers have responded to rapidly changing marketplaces and tougher competition with a variety of changes in the design process. CAD is almost assumed. Manufacturers are now looking for the management changes and technologies that will enable them to make the next jump in being able to design better products faster.

Management changes include **concurrent engineering,** in which manufacturing staff work with design staff from the outset of a design project. In the traditional approach, a design was completed and then presented to the manufacturing staff. The problem with this approach is that discovery of manufacturing problems associated with that design would require either an expensive redesign or expensive manufacturing accommodations.

One of the latest management perspectives is directed at establishment of the **agile enterprise.** The agile enterprise is one that, by design, can provide the rapid and often radical changes in products to meet the equally rapid and radical changes in the marketplace. Computer-based design and manufacturing technologies and, increasingly, knowlege-based systems help provide such agility. They do so not only by streamlining design and manufacturing processes, but also by integrating tasks and organizations.

For a product to be designed rapidly and well, the design process must be closely linked to marketing and manufacturing. Marketing must initiate the new product specifications that drive the design. Manufacturing staff must interact with designers through the concurrent engineering process to ensure that cost and quality objectives are best met. This requires new management perspectives and often requires new organizational structures. But technology also plays a key role. Upcoming chapters detail the ways that CAD and knowledge-based systems have risen to the occasion.

CAD technology has matured in the usual way: the technology improved, prices dropped, benefits became more focused, and the approach was validated. CAD is no longer sold primarily as a labor-saving tool, even though there are applications for which this is a realistic expectation. Instead, users report benefits of faster modification of existing designs and easier storage and distribution of completed designs.

Design changes are easily accomplished without starting from scratch. The computer-based design record is instantly accessible anywhere, over a computer network or over a telephone line. These changes are instantly available to designers, manufacturing staff, purchasers, planners, and schedulers at the home office and at remote plants. Vendors acquire part designs electronically from the original equipment manufacturers (OEMs) that they supply. These capabilities are critical as manufacturers work faster, smarter, and in closer contact with suppliers and customers.

Design technologies and products continue to evolve. The following information will soon be obsolete as the performance of computer-based products continues to increase, and prices continue to drop. However, this information is presented to provide a historical perspective on design capabilities. Many manufacturers, particularly small and medium-sized organizations, are challenged to keep track of these trends. The need for manufacturers of all sizes to become fully competitive in design and manufacturing technologies has led to the creation of a number of not-for-profit technology transfer organizations to help manufacturers understand and acquire technology.

The Great Lakes Manufacturing Technology Center (GLMTC) of Cleveland, Ohio, is one of the several Manufacturing Technology Centers established by the National Institute of Standards and Technology (NIST) and is affiliated with the nationally recognized Cleveland Advanced Manufacturing Program (CAMP). Mr. Gary Conkol, a staff member of GLMTC, helps manufacturing select and upgrade CAD systems. He sees current CAD capabilities and pricing falling into three groupings.

A typical personal computer (PC) based low-end system would include a 486DX66 PC, a 16″ color SVGA monitor, a tape backup device, and basic design software, and it would cost approximately $5,000 per user. Each design group will need to share a printer and an inexpensive plotter totaling $5,000. Thus, the so-called first seat will cost $10,000.

On such a system, software is available to support mechanical design, such as finite element modeling and kinematics as well as specialized applications such as computational fluid dynamics, electromagnetic fields problems, and acoustics. The computational speed will be slower and the display resolution will be less than workstation-based systems.

At the high-end, CAD systems are based on workstations such as Sun SPARCstations, Hewlett-Packard 700 Series, and others. Workstations priced at $20,000 are typically used with comparably priced software. A printer and a plotter totaling approximately $10,000 would be appropriate in this case. Thus, the first seat of such a system would total $50,000. This system will support more compute-intensive software, including advanced solids modeling packages as would be used to explore, for example, the dynamic performance of a part.

Between these two price points are systems costing approximately $30,000 for the first seat. These midpriced systems provide somewhat less speed and analytical capabilities than the high-end systems. Nonetheless, these systems provide many of the 3-D modeling capabilities of the high-end systems, enabling the user to judge the final appearance of the design in progress and to determine potential interference with adjacent parts. These modeling packages also provide mathematical descriptions of part surfaces, such as those required to generate numerical control (NC) tool paths and rapid prototyping equipment inputs.

Advanced CAD software packages now support a number of other features to streamline design. Parametric design capabilities let the user write equations to define features or relate features to one another. For example, one hole might be defined as always 1.56″ away from a certain surface. If the part is lengthened in that direction, the hole would automatically move along with the surface.

Another advanced functionality is **feature-based design.** It takes parametric design one step further, enabling the designer to define and work directly with features such as holes, slots, and ribs, rather than dealing with the individual points and geometries that make up the features. The designer manipulates features of interest and lets the CAD system handle more of the low-level work.

These are the boundaries of traditional CAD technologies. The systems provide powerful analytical tools to support the creation and manipulation of geometries. There are, of course, limits to CAD capabilities. A fundamental limitation is that the engineer or designer must bring all knowledge and insight to the CAD system. What is needed is a *smart* CAD system. Knowledge-based systems have demonstrated significant capabilities in this area and thus can be viewed as an element of next-generation CAD.

2.4 THE FUTURE

Tougher competition and heightened customer expectations force manufacturers to pursue every opportunity to develop products that are better, less expensive, and of higher quality. The design process is receiving scrutiny as an area in which improvements are possible. Manufacturers must increasingly view the design task in the broader perspective of finding the optimal design, the combination of geometries, materials, and processes that best meets customer and corporate goals.

The objective of optimal design is compelling. Many new materials and processes can provide significant cost savings or performance benefits, even for mature products. The execution is challenging. It is difficult to discover the relationships among geometry, materials, and processes that make best use of new materials and processes.

The new materials to be considered are often outside of the area of expertise for in-house staff. Designers are uncomfortable exploring new materials and processes for fear of overlooking new-to-them design or fabrication issues. Specific expertise, as well as a comfort level, may be lacking and inhibit exploration of opportunities.

Managing the interaction among design, materials, and fabrication is a further challenge because the underlying relationships are rarely known to the level of detail required to develop a truly optimal design. Anecdotal evidence may exist to provide some rough design rules of thumb. But to truly optimize across geometry, materials, and processing requires an explicit understanding of the interrelationships among the three. In short, a new type of model is required to capture the relationship among these different but related aspects of product design.

The latest CAD software is beginning to address the topic of optimal design. The designer may state design objectives, as related to part geometry, and the CAD system will vary geometric parameters to best meet these objectives. The future of CAD includes further progress in this direction, to enable designers to state broader objectives and enable the CAD system to optimize over design variables beyond just geometry.

Current CAD systems look to the designer for knowledge about design options, limitations imposed by geometry, materials, fabrication, and all the engineering details that separate successful designs from unsuccessful ones. Another element of future CAD will be built-in knowledge. This knowledge will provide advice, identify risky approaches, and suggest alternatives.

In fact, knowledge-based systems have demonstrated these elements of future CAD. Expert systems have captured the expertise of the best designers, as well as the design rules and recommendations of these experts. Neural net technologies have demonstrated capabilities that go beyond the capture of existing expertise, discovering previously unknown relationships. These are signficant capabilities that have a profound impact on design opportunities.

A commercial knowledge-based system combining neural nets, expert systems, and fuzzy logic, as described in upcoming chapters, integrates design and manufacturing by finding the product design and process parameters that, together, provide the best overall combination of product performance, quality, and cost. By integrating over the process domain as well as product design, this tool accommodates manufacturing constraints and balances raw material costs against processing costs, and both of these against product quality and function. This tool represents an interesting advance in information management, because the modeling and optimization is based on relationships derived from data and knowledge that were within the user's organization but not used to the extent now possible.

In upcoming chapters, knowledge-based system application examples will be presented for the design of discrete parts and products produced by batch and continuous process products. These examples demonstrate that, current CAD capabilities notwithstanding, significant new design opportunities now exist.

Similar opportunities exist in manufacturing. The next chapter outlines current computer-aided manufacturing capabilities and the manufacturing opportunities provided by knowledge-based systems. Succeeding chapters detail how three knowledge-based technologies—neural nets, fuzzy logic, and expert systems—provide better functionality, lower cost, and higher quality.

2.5 REFERENCES

1. Conkol, Gary K. "Initial CAD/CAM System Selection and Evaluation." Technical Paper No. MS92–336. Dearborn, MI: Society of Manufacturing Engineers, 1992.
2. Conkol, Gary K. "The Role of CAD/CAM in CIM-Part II." Technical Paper No. MS91–442. Dearborn, MI: Society of Manufacturing Engineers, 1993.
3. Shunk, Dan L. *Integrated Process Design and Development*. Homewood, IL: Business One Irwin, 1992.

Chapter Three

Computer-Aided Manufacturing:
Faster Response, Lower Costs

3.1 INTRODUCTION

Manufacturing technologies have evolved when two conditions are met: (1) a new technology enables an improvement and (2) this improvement addresses a perceived need. Improvements in design technology assist the manufacturer but are often invisible to the customer. In contrast, advances in manufacturing technology are often of great interest to customers because they result in products with better function, higher precision, more features, or lower cost. This customer *pull* has driven a steady succession of new technologies.

Computers have become a key manufacturing technology, and many manufacturers employ computer-aided manufacturing (CAM) at some level. Most manufacturers use computer-based manufacturing planning and management tools. Many use computers to monitor or control individual processes. The most progressive manufacturers employ computers to monitor and control every process and link processes and activities throughout the organization with a hierarchical data acquisition and control network spanning the factory floor and the organization as a whole.

These plantwide systems provide a critical information management capability. Operators and managers can track results and spot problems on the basis of data that is useful in their area but generated in another. The good news is that more information is available. The bad news is that the volume of data can be overwhelming.

One challenge is that the rate at which data are generated is outstripping users' abilities to process those data. Another challenge is that the nature of the data is changing and becoming more difficult to interpret as manufacturing processes and controls become more sophisticated. CAM has succeeded in providing more data, but not necessarily more insight. Better data interpretation and decision support capabilities are needed.

This chapter outlines manufacturing technology trends, with a focus on computer-based technologies. Knowledge-based systems are shown to extend CAM in directions that reflect current manufacturing needs for training, decision support, and process optimization.

3.2 THE PAST

Before the time of computers, one could say that the first manufacturing technologies produced stone tools. Manufacturing came of age in the Industrial Revolution, as advances in science were adapted to practical applications. Running water and burning coal produced power that ran machines. These machines provided new products and lifestyles. An agrarian society became an urban society. Manufacturing created wealth and a new middle class. Standards of living rose as what were formerly luxuries became affordable.

Automobiles were a rich man's toy when they were built by hand and replacement parts had to be hand-fitted. Cars became affordable, and widely accepted, when manufacturing capabilities provided low-cost and interchangeable parts.

American manufacturing flourished from the late 1800s to the middle 1900s. The rules were different than they are today. The emphasis was on low costs, and it was a very successful strategy. Low costs were achieved by ever-increasing production rates. As costs dropped, markets grew. Competition was minimal by today's standards. Long production runs helped hold down costs and were quite satisfactory in an era of slower changes in technology and customer expectations.

The long production runs also minimized the effects of rudimentary design and quality control technologies. Quality was achieved by inspection. The production equipment was repeatable if not accurate. By the time a few hundred thousand copies of a part had been run, the machine was adjusted so that the mean values of any critical dimensions matched the target values pretty well. Overall, it was a workable scheme. In fact, it was an excellent scheme at the time because it met the needs of the marketplace.

Until the 1960s, this was the winning formula:

- Process and equipment improvements focused on higher production rates.
- Processes were designed and factories were arranged to provide maximum equipment utilization.
- Direct labor costs were over half of total direct costs, excluding the cost of raw materials.
- Management was top down; labor provided muscle.
- Labor was widely available; apprenticeship programs provided the manual skills required, and technical skill requirements were minimal.
- Design, analysis, planning, and record keeping were paper-based and manual.

And then, the rules changed. New competition emerged. This new competition employed new technologies. The formerly winning formula, the old management approaches, and the older manufacturing facilities started losing ground. Market shares diminished, and entire industries vanished.

3.3 THE PRESENT

For manufacturers, the present environment is a very different place. Worldwide competition and new customer expectations require new manufacturing technologies and new ways of doing business. New processes are required to improve quality and reduce costs.

Other problems have less to do with technology and more to do with people. Worker skills have not kept up with evolving requirements. Skilled trades positions are unfilled, even in times of high unemployment. The absence of apprenticeship programs makes it difficult to replace skilled tradespeople. Increasingly sophisticated manufacturing processes are required, but operators exhibit decreasing skills in basic mathematics and problem solving. Some experts estimate functional illiteracy among current American factory workers to be 20 to 30 percent!

Progressive manufacturers are addressing skills issues with in-house employee training programs. Courses include basic reading, mathematics, and the problem-solving skills required to operate equipment. Manufacturers are increasingly taking advantage of eyes, ears, and brains on the factory floor by training workers to assume broader process monitoring responsibilities.

Programs such as **total quality management (TQM)** and **continuous improvement** reflect the need to improve quality and the use of resources—human and otherwise. These programs call upon the people on the factory floor who have first hand knowledge of many quality and operational problems that will inevitably escape the notice of management. Implementing these programs often involves challenges in motivation and trust building. Some see these programs as the latest management gimmick that is mostly about getting more work for the same wages. But successful manufacturers act upon the realization that even as direct labor costs drop to a smaller fraction of total costs, this resource must be better utilized.

This goal is also reflected in employee empowerment programs and improvements in the work environment. Empowerment programs strive to raise among workers a feeling of ownership of the problems and solutions in their work areas. More effective use of people, as well as more interesting work environments are achieved with new factory layouts.

The traditional factory layout called for similar equipment to be grouped together. For example, all of the grinders were together, as were the milling machines. One operator ran one or more of one type of machine. Work in process might be transferred a considerable distance from one work area to another and might crisscross the plant in the process.

The new strategy is to arrange equipment in work cells grouped by product type. Work in process moves short distances from machine to machine within the cell, and so material handling is minimized. One person is responsible for the several types of machines in the cell. The operator's duties are more varied and therefore more interesting, encompassing more of the tasks associated with that manufacturing process. The cell layout provides greater direct labor utilization but lower machine utilization. The operator moves from machine to machine in the cell, and the process is set up to keep the operator busy even if individual machines sit idle after finishing their cycle.

The following examples illustrate the use of computer by progressive manufacturers.

Kennametal's Solon, Ohio, plant manufactures tool holders and fixturing for their line of cutting tools. The clean, bright, and colorful environment incorporates a work cell layout. The Kennametal plant also illustrates the impact of computers on manufacturing. Sophisticated CAD systems, customized to meet Kennametal's needs, support design requirements. Computers throughout the manufacturing area are networked to plantwide systems and display operational status and scheduling information. Flexible manufacturing and other computerized equipment is employed to provide rapid machine setups. Older equipment is integrated into the floor layout and is used as required.

Ford's Avon Lake, Ohio, plant and the Honda Marysville, Ohio, plant also illustrate the impact of management and technology changes in automotive manufacturing. The Ford plant has built the Mercury Villager and Nissan Quest minivans since 1992 in a new facility adjacent to an older plant that builds the larger Econoline vans. The Honda plant has built Accords since its opening in 1982. Although specifics differ, similarities between the plants dominate.

On-site training centers teach employees basic mathematics and problem-solving skills, as well as manufacturing-related skills such as **statistical process control (SPC).** Workers learn to measure production variables and from these calculate and plot simple statistical measures of process trends and capabilities. In this way they can anticipate trouble and help monitor and control quality.

Evidence of computers at work is seen in the smoothly-functioning **just-in-time (JIT)** systems. Component suppliers receive orders by electronic transfer from the assembly plants. The orders may specify multiple shipments per day. Tentative long-term schedules are revised through the computer hook-up to reflect needs that become more precisely known as the shipment data approaches. Suppliers ship product based on these, with trucks loaded in reverse order from production requirements.

At the Honda plant, material is received on customized pallets and moves in lock step from trucks to conveyors. Hours, not days or weeks, of inventory are maintained. Pallets are returned to the loading docks, and other material handling duties are handled by automatic guided vehicles (AGVs) that follow marked paths and subtly make their presence known by playing music (including Christmas tunes during the season!)

The Kennametal, Ford, and Honda plants demonstrate the current state of the art in process monitoring and plantwide computer networks. Large monitors around the factory floor display the status of individual machines, cells, and assembly or process lines. The queue of upcoming orders is also displayed. The monitors display alarm messages, identifying problems with processes or individual machine parts, or other problems requiring immediate attention.

Plantwide computer networks support hierarchical control. At the lowest level, **programmable logic controllers (PLCs)** and **distributed control systems (DCSs)** monitor and control individual devices and processes. They identify alarm conditions by comparing current values of significant process parameters to upper and lower limits that are predetermined to represent the bounds of acceptable behavior. If these limits are exceeded, an alarm signal is typically displayed at the controller that identified the fault and transmitted to the plantwide network.

Moving up the control hierarchy, computers provide supervisory control of processes and areas. For example, each manufacturing cell has a controller that monitors the individual machine controllers within the cell. Additional layers of control hierarchy may be used as appropriate, depending on the size and diversity of the manufacturing activity. At the highest level, the plantwide system coordinates manufacturing operations with other activities.

The hierarchical control network is a coordination and record-keeping tool. It is also a real-time process monitoring tool. The plantwide indication of alarm conditions helps in summoning maintenance staff who may be a 10-minute walk away. However, users report that further alarm processing capabilities are needed.

The unmet needs in alarming capabilities can be characterized as data acquisition and interpretation. One challenge is to provide an alarm to cover every possible failure. Another challenge is to interpret alarms to minimize distracting nuisance alarms and to provide useful troubleshooting and repair instructions.

Meeting the first challenge requires the process controller both to understand and to detect all possible failure modes. Many failure modes are obvious and easily detected. Other failure modes are difficult to detect. Some occur within a process or a device and cannot be detected by external sensors. Others involve changes in process parameters for which sensors are not available. Still others involve undesirable interactions between several parameters. These interactions may be inadvertently overlooked because their frequency of occurrence is so low. Or, as a practical matter, the engineering effort required to identify the many possible interactions and develop an appropriate alarm strategy may be too time consuming.

An example of the time required to develop an alarm system is provided by an innovative assembly machine built by the General Electric Lighting Business Group in the early 1980s. This machine feeds, forms, assembles, and welds the wire and sheet metal parts that make up the inner structure of a line of high-efficiency discharge lamps. It uses a programmable logic controller (PLC) to drive the hundreds of air cylinders, welders, bowl feeders, and other devices on the complex 24-station machine.

A sophisticated alarm system in the PLC identifies jammed mechanisms, absence of parts, and faulty welds. Alarm messages are displayed on an alphanumeric panel, so that operators or machine adjusters can replenish parts, clear jams, service welders, or diagnose more involved problems. The alarm system monitors the speeds and sequencing of moving parts, as detected by position sensors. Alarm messages are triggered by rules that identify deviations from the proper sequencing or timing tolerances of the various assembly processes.

This alarm system plays a vital role in commissioning and maintaining this large, complex machine. The price that had to be paid for this convenience, beyond the extra PLC and display hardware to support the alarm system, was the need for an additional six months of the control engineer's time to develop this alarm system beyond that required for him to develop the rest of the control system. This time frame would have been even longer if the alarm system had not been developed by the engineer who was intimately familiar with the machine and its controls.

Six months of development notwithstanding, this alarm system manages a process simpler than many. The assembly machine consists of 24 relatively independent stations. Interactions between devices and processes exist within each station, but station-to-station interactions are minimal. This means that the number of possible alarm conditions is far lower than it would be with a more interdependent process.

As a practical matter, alarm systems generally do not incorporate knowledge of these interactions. The result is a problem known as *nuisance alarms*. One process deviation or equipment failure may cause a number of downstream process parameters to change. The effect will be that a single root cause will trigger many symptoms. Since the alarm system does not understand these interactions, each symptom will trip an alarm. Thus, a single event may be announced by an overwhelming barrage of alarms, only one of which represents the underlying problem. But which one? In complex interconnected systems such as nuclear power plants, a single fault may trigger literally hundreds of alarms. The problem-solving task becomes one of deduction, to determine what single cause can explain the multiple effects.

A need exists to provide better alarm processing capabilities in several areas. Alarm system development is currently too time consuming for all but the simplest processes. But the need for effective alarm processing is growing. The trends in manufacturing are in the direction of increasingly sophisticated processes and shrinking (and perhaps less skilled) maintenance staffs. For this reason, manufacturers need to minimize nuisance alarms that are not only distracting but also significant impediments to problem solving. Beyond this, alarm processing must move beyond mere identification of problems to identification of solutions.

3.4 THE FUTURE

Control system users have identified the following technology-related requirements to meet the needs that are imposed by tougher competition, shrinking staffs, and the need for more effective problem-solving.

- More effective monitoring and diagnostic systems and better ability to predict problems are required to provide more equipment uptime.
- Operators want more focused advice, including computer-based reasoning regarding current problems, better

filtering of nuisance alarms, and presentation of recommended actions. These recommendations could be based on records of solutions to both prior problems and simulated situations. Recommendations may also be ranked on the basis of need for immediate response and certainty of correct diagnosis.

- Better user interfaces are needed to present more information more clearly and to minimize the skill required to interpret and act upon the information presented.

Knowledge-based systems have demonstrated capabilities in these areas. Upcoming chapters describe the ways that knowledge-based systems monitor processes, interpret data, and present action items in ranked order. Knowledge-based technologies are also providing significant new capabilities in manufacturing management.

Computers took on a new role in manufacturing management with the introduction in 1965 of **material requirement planning (MRP)** systems. These systems generate a material requirements list in response to given production requirements. In this way, inventory management, purchasing, and shipping activities are linked to manufacturing. In 1979, **MRP II (manufacturing resource planning)** systems were introduced. These systems take the next step, identifying the facilities and equipment as well as the raw materials required to meet given production requirements. This additional information supports scheduling and capacity planning by production planners who must determine how to meet processing and delivery requirements with available facilities.

An emerging market is developing for products that expand the scope of manufacturing support systems to encompass the entire organization. A term such as MRP III would not convey the broader scope of these systems, and so new buzzwords are used to describe these new products. Among them are **enterprise resource planning (ERP), customer-oriented manufacturing management systems (COMMS), and manufacturing execution systems (MES).** By integrating data and software tools from marketing, manufacturing, sales, finance, and distribution, these systems strive to move beyond optimizing production alone to optimizing the organization's many (and sometimes conflicting) requirements of low cost, rapid delivery, high quality, and customer satisfaction.

These new products address needs similar to those identified in the areas of production monitoring and control. The many computer-based tools and databases now available across plantwide networks need to be better integrated. Better methods to interpret and present information are also required. These goals will be achieved with new database and knowledge-based technologies.

Looking at the bigger picture, some say that the future will be the Age of Information and that effective use of information will be the critical element of future success. The acquisition and interpretation of knowledge on an organization-wide scale will become an issue of strategic importance, and the capabilities of knowledge-based systems will provide a strategic opportunity. Technologies and management approaches will focus on the acquisition and use of information from internal and external sources. Knowledge-based systems will play an increasing role, interpreting data, solving problems, and identifying successful responses to new needs and opportunities.

Upcoming chapters describe ways that knowledge-based systems address these strategic issues, particularly in the context of design and manufacturing. The next chapter describes the nature and strengths of three distinct but complementary knowledge-based technologies: neural nets, fuzzy logic, and expert systems.

3.5 REFERENCES

1. Babb, Michael. "What DOS Operators Want." *Control Engineering,* November 1991, p. 53.
2. Connor, Susan. "Integrating Quality Efforts in Process Production." *Manufacturing Systems,* March 1992, pp. 35–40.
3. Gage, Steven. "What Is Happening in Manufacturing?" *Works in Progress,* Winter 1992/1993, pp. 1–3.
4. Honda 1992 Fact Card. Marysville, OH: Honda of America Mfg., Inc.
5. Inglesby, Tom. "Industry Profile: An Interview with Larry Ford." *Manufacturing Systems,* October 1992, pp. 34–38.
6. Melnyk, Steven A., and Ram Narasimhan. *Computer Integrated Manufacturing: Guidelines and Applications from Industrial Leaders.* Homewood, IL: Business One Irwin, 1992.
7. Metz, Sandy. "Making Manufacturing Better, Not Just Faster." *Managing Automation,* August 1990, pp. 22–24.
8. Nagel, Roger N. "The Agile Opportunity." *Works in Progress,* Winter 1992/1993, pp. 10, 15.

Chapter Four

Knowledge-Based Technologies:
New Ways to Acquire and Apply Knowledge

4.1 INTRODUCTION

Progressive manufacturers have recognized the importance of information management and that better management of information will require new approaches and new technologies. This chapter outlines some issues and opportunities regarding the management of information in support of design and manufacturing. The issues concern challenges in collecting, organizing, and applying information. The opportunities, as presented here, are based on new ways to apply computers using the advanced computer technologies of **expert systems, fuzzy logic,** and **neural networks.**

These technologies have demonstrated clear benefits in support of design and manufacturing tasks. The chapter presents the different yet complementary capabilities of each and the requirements for their successful application. In following chapters, applications of these technologies are presented to clarify the opportunities they present.

The advanced computer-based information management technologies called **knowledge-based systems (KBS)** share a common goal: better acquisition and application of knowledge. Knowledge is required to design products and processes that are feasible, cost-effective, and functional. Once the product is designed and the manufacturing process is under way, knowledge is used to analyze data, identify trends and problems, and define appropriate corrective actions.

There are several distinct technologies within the category of knowledge-based systems, and they meet the needs of manufacturers in different ways. Expert systems and fuzzy logic incorporate knowledge in the form of rules. Neural nets actually discover knowledge in the form of relationships hidden within data. These technologies have provided manufacturers new yet proven tools, each with certain strengths and limitations.

The strengths relate to the types of knowledge that may be captured and the types of decision support provided. The limitations relate to the resources and types of knowledge required to build a successful system. To identify applications and select tools, manufacturers should consider the nature of their applications and the information available to support those applications.

4.2 THE NATURE OF INFORMATION

Information may be classified as numeric or symbolic. Numeric information is expressed as numbers, and these numbers are easily manipulated by people and computers using the unambiguous rules of mathematics and various analytical methods and software tools.

Symbolic information is expressed with symbols. Common symbols include the names of places and things and the other words that make up spoken languages. These symbols in the form of words have an arbitrary but commonly agreed-upon usage. Meaning is conveyed and knowledge is expressed by collections of symbols in the form of words and sentences. Much design and manufacturing information is symbolic. Design guidelines and process setup and maintenance procedures are largely word-based.

Computer-based systems have been far more successful dealing with numeric information. Precise, unambiguous rules exist to manipulate and interpret numeric data. These rules make up the various disciplines of mathematics and numerical methods, and they tell the computer what numbers mean and what to do with those numbers.

But the computer's ability to manipulate symbolic information is limited by the ambiguous, context-dependent nature of much symbolic information. Since traditional computer technologies require specific, unambiguous instructions, a given interpretation of symbolic information will be appropriate only for the circumstances that were envisioned in advance and then explicitly programmed.

Consider: "He hit me" and "He scored a hit." In one case "hit" is a verb, in the other—a noun. Furthermore, the sentences are meaningful only in particular contexts that are not revealed here. One might assume that the first sentence refers to human interaction and the second to baseball. But a computer would have to be explicitly programmed to be able to act upon those assumptions and correctly interpret the sentences.

A further challenge in computer-based systems is that they cannot mimic a capability that in humans is called *common sense*. Since computers cannot make assumptions that would be obvious to humans, computer-based systems can come to ridiculous or trivial conclusions unless explicitly guided. In theory, this guidance would include a sense of environment and context to mimic that which people automatically acquire through their interpretation of inputs received from their senses of sight and hearing and from their understanding based on life experiences. In practice, with current technology, this guidance can be provided only in the form of explicitly envisioned and programmed circumstances. This means that computer-based systems have very definite, if not always apparent, limits.

Computer-based systems must also deal with real-life ambiguity in the form of uncertainty. This uncertainty can be in two forms. The first is one of perception. One may wonder, is that really a long-lost friend on the other side of the restaurant? In the same way, one must deal with uncertainty in the accuracy of data and sensor readings. For example, a process variable may be measured as 5. But is the sensor calibrated well enough to know that the true value is not 4 or 6?

Statistics deals explicitly with uncertainty in data. Rules are provided to enable users to establish a confidence level for conclusions based on a certain set of data, depending on the size of the data set and the methods used to collect that data. The uncertainty cannot be eliminated, but its magnitude can be estimated.

The second type of uncertainty relates to the analysis or action that is most appropriate under given circumstances. One might wonder if that long-lost friend on the other side of the restaurant should be acknowledged, considering the last meeting and the many years that have elapsed. In the same way, there may be uncertainty in the way to deal with a situation that is not fully understood because it is described by uncertain data. For example, if a process variable is now 5 and it should be $5^1/_2$, the course of action is clear. But if the process variable is not accurately measured, and may be anywhere between 4 and 6, the appropriate action is less obvious.

To be effective problem-solving tools, knowledge-based systems must deal with these problems of uncertainty. Another factor that must be considered is that of knowledge representation. Knowledge must be properly represented, that is, captured and delivered in a form and depth useful to users. Terminology must be correct. Beyond this, the structure of knowledge incorporated in the KBS must match that of the real-world problem. Each problem domain involves specific types of knowledge, with specific relationships among these types of knowledge. An expert understands these relationships, at least intuitively. A KBS must reflect this structure of knowledge if it is to be viewed as an efficient problem-solving tool worthy of use.

For example, a KBS might be developed to diagnose a car that won't start. A proper KBS design must recognize that causes of not starting fall into several broad categories (won't crank, no spark, no fuel) and that within each of these categories are further details, also in a hierarchical structure. The KBS should reflect the structure of the problem in that, for example, it would know that spark plugs are part of the ignition system, so that if a prior diagnostic test identified the problem to be fuel related, the KBS would not pursue spark plug problems. Instead, the KBS would pursue symptoms and potential diagnoses related only to the availability and delivery of fuel.

The point is that the design of a KBS must, like any other sophisticated design project, begin with an analysis phase. This analysis will reveal, among other things, the structure of the problem to be solved and the nature of the information available to solve it. The analysis phase will help the developer gain a clearer understanding of the problem-solving challenge and the best solution. The results of this analysis may not be explicitly visible in the finished system, but the absence of this up-front analysis is usually evident in a system that is too limited or too hard to use to merit ongoing support.

Of course, other aspects of KBS implementation will determine the ultimate success of the system, as will be covered in upcoming sections. Before delving into these specifics, it may be useful to understand how knowledge-based technologies relate to other advanced computer technologies.

4.3 ARTIFICIAL INTELLIGENCE

The term *artificial intelligence* (AI) conjures up many images, not all of them positive. To some, AI is "weird science." Some think that AI is the basis on which computers will rule the earth, following the lead of HAL, the renegade computer that took command of the spaceship in the science-fiction movie *2001: A Space Odyssey.*

Others see AI as computer-based magic, along the lines of the ancient Greek oracle. Ask the all-knowing computer a question and it will give you an insightful answer.

Still others see AI as enabling computers to take on human capabilities in a more benign way. In 1950, the British mathematician Alan Turing described the test of computer-based intelligence that bears his name. Computer-based intelligence would be demonstrated if a person could carry on a conversation with a computer and not realize that the conversational partner was not human.

Human-to-computer conversations have been demonstrated and a dialog maintained at some level. The conversation was one-sided in the way that a psychologist might interact with a client. For each statement made by the client, the computer would respond with a statement that largely paraphrased the original statement and sought to move ever so slightly beyond the original statement. The effect was intriguing to some, but it was evident that the computer responded only to the explicit cues presented to it and only according to preprogrammed associations and rules of response.

What has not been demonstrated is anything remotely resembling the level of common sense or creativity that enable human conversations to be truly interactive and to proceed in unanticipated directions. Some wonder whether computers can ever incorporate intelligence as defined by the Turing test. The skepticism with which AI is viewed in some circles is due in part to AI's inability to meet very aggressive goals such as this. And, AI is true to the form of other new technologies in that proven benefits turn out to be a good bit less sweeping than those originally claimed.

Overblown claims notwithstanding, AI technologies have demonstrated real benefits in a wide variety of applications. These successes fall under a more common and more limited definition of AI—that it enables computers to solve types of problems for which they were traditionally not well suited. This covers a number of areas, none of which include magic or human levels of intelligence.

One of these areas is **pattern recognition.** The underlying premise of pattern recognition is that some tasks are better performed by analyzing the entire pattern of, for example, a visual image rather than looking at individual elements of that image. The point is that similarities are more easily recognized without distraction by irrelevant differences. AI—in particular, neural net-based pattern recognition—has also demonstrated significant benefits in speed and accuracy.

Manufacturing applications of pattern recognition include **computer vision**-based inspection. The following reported benefits have been compared with traditional image analysis techniques. Neural net-based systems provide faster analysis, so they can either provide current levels of inspection faster and thus enable faster assembly-line speeds, or they can perform more sophisticated analysis in the same amount of time. An enhanced ability has also been reported to distinguish those defects considered to be significant and cause for rejection of the part from those deviations from the ideal part that are of no consequence and thus can be ignored.

Neural net-based **signal analysis** has demonstrated significant capabilities in detection and recognition of extremely weak signals. In a defense-related sonar signal analysis application, the system detected signals hundreds of decibels below noise and other distracting signals, providing capabilities not possible or practical by other methods.

Robotics is often considered a branch of AI, although many consider robots to be merely general-purpose machines that may or may not incorporate AI. As general-purpose machines, robots have succeeded in applications such as welding and spray painting. Their mechanical configuration as arms with many degrees of freedom enables them to provide the contortions that characterize particular tasks that are widely used, for example, to manufacture cars and appliances. Their ability to switch immediately from one preprogrammed set of motions to another meets the need in these industries to accommodate several different product types on one assembly line.

Their assumed association with AI notwithstanding, one of the primary limitations of robots has been their absence of intelligence, artificial or otherwise. Robots are completely stupid. Their knowledge of their environment is derived from sensors and perhaps a computer vision system. Sensors detect specific anticipated conditions. But they don't provide any sense of environment or, more important, changes in environment. Computer vision systems can provide a more comprehensive view of the world, but they are constrained by the limited image analysis that is possible at assembly-line speeds.

Practical robotic applications will increase when robots become smarter with better perceptive and analytical capabilities. Increased perception through more and better sensors will provide a more comprehensive sense of environment. Improved problem-solving capabilities will enable robots to interpret changes in the environment and in this way adapt more readily to those changes that are benign but have traditionally required attention by scarce and expensive robotic engineers. An example of a benign change is when parts to be picked vary slightly in shape or come to rest in the pickup location in a slightly different place or orientation. Knowledge-based computer vision and robot control systems are expected to enable these capabilities to interpret and adapt to environmental changes.

Knowledge-based systems, the focus of this book, are a part of AI. These systems are typically focused on practical decision support applications in areas such as retail management and financial applications, as well as design and manufacturing support. They incorporate knowledge about the performance of a task or analysis of a situation. In one approach, this expertise is acquired from human experts in the form of rules and is incorporated in so-called expert systems. A variation on this approach called fuzzy logic provides a rule-based but more intuitive way for developers to express control objectives. Neural networks, a third knowledge-based system technology, enable discovery of relationships within

FIGURE 4-1
Knowledge-Based Technologies

	Expert Systems	*Fuzzy Logic*	*Neural Nets*
Manufacturing applications	• Product design and configuration • Equipment maintenance • Training	• Process control • Embedded controllers for cars and appliances	• Product design • Process optimization • Inspection and quality control
Source of knowledge	• Human expert	• Human expert	• Data
Strengths	• Provides precise answers • Explains reasoning process • Inherently symbolic	• Easier programming • Accommodates qualitative inputs	• No programming • Unique modeling capabilities • Adaptive, nonlinear
Limitations	• Must know unambiguous rules • Difficult to develop and maintain	• Narrow application area • Must know rules	• Requires adequate data • Inherently numeric

data and can thus discover knowledge and in this way adaptively improve, even in a changing environment.

These three knowledge-based technologies are similar in that knowledge is acquired from people or through discovery. In contrast, other AI applications such as speech or character recognition might not be considered knowledge-based systems if their generic capabilities are inherent in the system rather than application-specific by design or discovery.

The many successful KBS applications in design and manufacturing have validated these technologies with down-to-earth solutions of common problems that contrast with the image of AI as perhaps an interesting research topic, but certainly not a practical problem-solving technique. The three KBS technologies are distinctly different in nature, and each has particular application strengths and limitations. These attributes are summarized in Figure 4–1 and detailed in the following sections.

4.4 EXPERT SYSTEMS

Expert systems are the most established and recognized knowledge-based technology. In fact, many consider the term artificial intelligence to be synonymous with expert systems. Expert systems capture human problem-solving expertise in the form of rules. This approach is useful for problems that are difficult enough to require special training or experience and for which an expert is available to identify concise and complete problem-solving rules. Ideally, the expert has sufficient depth of experience to be able to look past the obvious and provide insight in the form of *hunches* that provide problem-identification shortcuts.

An expert system user poses a problem or a situation to the expert system. The expert system then calls upon its internal understanding of the logic and issues pertaining to that particular problem. A question-and-answer dialog between the expert system and the user enables the expert system to clarify the situation and ultimately provide the user a diagnosis or recommended next step.

The user may seek to better understand the problem-solving approach or perform a reality test on the answer by asking the system how it reached the given conclusion. The system will respond with the step-by-step reasoning process based on the specific rules used and the conclusions derived from these rules.

The system's logic can call upon either forward or backward chaining. In a forward chaining mode, the user defines a current or hypothetical situation. The system uses its internal knowledge, including rules acquired from the human expert, to determine the outcome that can be anticipated.

Backward chaining is the reverse. The end point, such as a current problem or symptom is presented. The system backward chains, or proceeds logically backward in time, based again on internal knowledge, to infer the most likely cause of the problem.

Expert system users are typically less experienced staff members. They can use the system not only for decision support but also for training. The system can be an effective training tool because users observe reasoning processes and applications of knowledge in specific contexts. The decision support role of expert systems is useful for both novices and experts. Novices draw upon the system's expert knowledge to handle more challenging problems than they would otherwise. Experts also benefit because they are available for the more challenging or unusual tasks.

An expert system contains three basic elements. The expert's knowledge is contained in the so-called knowledge base. The second part, the inference engine, contains the rules of logic and analysis that are applied to information in the knowledge base. The third part, the controller, manages the system, applying the inference engine functionality to the appropriate knowledge.

Commercial expert system *shells* or development systems provide the inference engine and system manager. When a shell is used, the expert system development task is reduced to development of the knowledge base. Even so, this is a large task often requiring at least two people—a knowledge engineer and a domain expert. The knowledge engineer extracts and structures knowledge acquired from one or more domain experts. This is a difficult and time-consuming task. The knowledge engineer must interact with the expert in a way that makes the expert comfortable and cooperative and willing to provide complete and concise information. The knowledge engineer must learn everything about a domain he or she typically knows nothing about.

The domain expert is critical to the development of an expert system (although less-desirable alternatives exist). The point of using a knowledge engineer, in addition to the clearly necessary domain expert, is to ensure that the domain expert remains focused and to ensure that rules of logic and good problem-solving techniques are used. The role of the knowledge engineer is to discover the structure of knowledge in the domain of interest, so that the resulting expert system will reflect this structure. The ideal case is to find one individual who is both a domain expert and a capable knowledge engineer. This eliminates the time-consuming interaction and the possibility of errors in communication between two or more individuals.

The domain expert has a challenging task. He or she will be asked to reduce years of experience to a list of rules. It is often difficult to remember all the many individual and slightly different problems and solutions that have occurred. It is harder yet to generalize these specifics into rules that are applicable over some range of circumstances. Many experts have difficulty identifying rules as such, preferring to specify only a particular response to a particular problem. The knowledge engineer must then help the domain expert abstract these experiences into rules.

The knowledge engineer asks the expert to describe the range of problems the system is to cover, how the problems are identified and categorized, and the logic behind the solutions. An expert's special knowledge often enables him or her to focus quickly on a particular problem without first going through every possible cause. This use of rules of thumb, called **heuristics,** limits the possibilities that must be searched. The knowledge engineer wants to identify and incorporate these heuristics because they are a critical element enabling an expert system to outperform what would otherwise be merely a computerized list of rules.

The knowledge engineer also needs to find the best representation and the appropriate depth of knowledge in a particular area. For example, the expert system must understand that "car won't start" probably has to do with the engine, and that engine knowledge is best divided into knowledge about the electrical and fuel systems. Starting a problem-solving dialog, the system might deal only with shallow knowledge, such as "the battery must be charged" and "there must be gas." But if the obvious problems are not found, the system must go to increasingly deeper knowledge. The next step might be "the ignition system includes six spark plugs, each of which must be installed and working." Going deeper yet might invoke the rule "for a brand X spark plug, the gap must be Y."

Many problem-solving domains are not so clear-cut. Uncertainty may be involved in the data that are used to identify the problem. This uncertainty may arise from sensors that are faulty or have not recently been calibrated. Errors in data may be caused by omission or transcription errors. Unqualified observers may provide errone-ous input, based on their errors in perception or interpretation.

Expert system developers need to be aware of these potential data errors. Many expert systems incorporate rules that provide data reality tests. In addition, system designers must consider the ways that uncertainty will propagate through the system, consider-ing the ways that uncertainty regarding individual data or rules will affect the overall conclusions reached by the system. To repeat the prior example, if the target value of a control variable is $5^1/_2$ and the current value is 5, a course of action is clear. If a sensor reading of 5 could indicate a true value of 4 to 6, the expert system must include a strategy and rules either to improve the data or to take action that will accommodate such uncertainty.

To acquire knowledge in sufficient depth and breadth, the knowl-edge engineer must be inquisitive, logical, personable, and perhaps persistent. The knowledge engineer also needs an expert who is available, cooperative, and truly expert. This point cannot be over-emphasized. Many expert systems have failed from lack of exper-tise. In some cases, when a true expert was not available, a textbook was used instead. In these cases, the expert system tends to be shallow or contains incomplete heuristics, seriously limiting the utility of the system. In other cases, experts see the expert system as a threat to their stature or continued employment, and so they do not fully support system development or maintenance.

One example of this was provided by Shearson American Ex-press. Its expert system, called K:Base, identified borrowers poten-tially interested in exchanging interest rates. Traders typically man-age this process known as interest rate swapping and generate commissions for themselves as part of fees that can approach $1 million per major deal. In 1984, it was reported that the recently completed K:Base had earned $1 million in two months of operation. But within a year the system was gone. Among the problems was that traders were called upon to keep K:Base current and therefore useful. Yet one trader generates higher commissions than the next one on the basis of his or her own closely guarded expertise. Traders saw no incentive to share their expertise. Since the intended users did not support the system by providing it their best and most current expertise, it failed.

A successful manufacturing example is described in an upcoming chapter. An expert system was developed by John Deere to support the diagnosis and maintenance of a new line of resistance welders being installed in one of its plants. This expert system incorporated expertise from a number of staff members. The electrician most familiar with the previous welders provided vital support to the development and verification of the expert system. He cooperated fully only after he was convinced that the system would not replace him or his peers but would instead help them do their jobs better. The lesson here is to consider carefully and manage the motivations of system developers and users.

Systems also fail because of lack of management support. This can include unrealistic expectations, poor planning, or inadequate management of the knowledge engineering, system development, or maintenance processes. Based on these problems, Michael Stock, president of Artificial Intelligence Technologies, estimated that only 10 percent of medium and large-scale expert systems are ultimately successful. This low percentage certainly calls into question the merits of the approach.

There are nevertheless many successful systems, and there are steps that can be taken to help ensure a successful expert system implementation. Managers must carefully define goals. They must secure the up-front support of users and experts. They must also provide sufficient resources to do the job right, including prototype and final system development, testing, user training, and system maintenance.

In the mid to late 1980s, most expert systems were programmed in languages developed specifically for AI. In the United States, LISP (for LISt Processing) was most popular. In Europe and Japan, Prolog dominated. These languages were developed specifically for symbolic and logic processing. In the early days of LISP-based expert systems, achieving reasonable system performance required a $100,000 LISP computer. Programming was additional and typically expensive because specialized, scarce LISP programming skills were required. In the days before expert system shells, systems were developed from scratch, including the user interface as well as the inference engine and system manager.

Since then, expert system development has become far easier as a result of the availability of expert system shells. These programming tools include a built-in inference engine, user interface, and programming environment. Programming is often done in an English-like language, that is, a language that uses English words in a specialized format. This makes system support easier, since LISP programmers are not required. Shells are available to run on PCs and general-purpose workstations.

The engineering workstation (typically running UNIX) and expert system shell to support a typical commercial caliber expert system might cost $10,000 to $30,000. However, system development still requires significant amounts of specialized expertise. A two-to-four-person–year project is typical for the development, installation, and training required for a medium-sized system of several hundred rules.

The chances for a successful implementation are also improved by picking the right problem. Some guidelines follow:

- An answer exists and can be found by an expert in minutes to hours based on rules that can be explicitly stated.
- The problem is neither trivial nor one of common sense.
- The problem is deep in a particular area rather than broad across a wide area.
- A clearly definable benefit can be identified, with interested and supportive users.
- The system does not require ongoing maintenance to remain current.

System maintenance issues are a major limitation of expert systems. It is often difficult for anyone other than the original developer to know how to update the system to reflect more than the most minimal changes in information or problem-solving approaches, and the original developer may no longer be available.

The original developer has a difficult-to-duplicate understanding of the structure of knowledge in the problem being addressed and of the way that structure is reflected in the expert system. This understanding is required to update the system to reflect new knowledge without inadvertently upsetting the structure, and therefore the usefulness, of the system.

Dependence on the original developer may also be based on the use of LISP or Prolog. For this reason, there is a strong and valid preference for expert systems developed with English-like languages.

The risk imposed by these dependencies can be minimized by maintaining access to the original developer. Maintaining access is easier if the developer is on staff rather than a consultant. However, not many organizations currently have or can justify the availability of such skill sets in house. For those that do, that staff member may not be available at the required time, because he or she may be involved in another project. The use of several staff members for system development may be justified purely on the basis of future availability for support.

None of these approaches fully overcomes maintenance issues as a limitation of expert systems. Although expert systems have demonstrated clear capabilities and benefits, issues such as development and maintenance costs have driven the development of alternative KBS technologies. Part of the appeal of fuzzy logic, as described below, is that in some applications there are fewer rules to write and maintain. Similarly, neural nets are appealing because there are no rules at all (although understanding of the problem is still required, as is access to relevant data).

4.5 FUZZY LOGIC

Fuzzy logic is an increasingly visible knowledge-based technology. It is based on fuzzy theory, first presented by Lofti Zadeh in 1965. After over 20 years of benign neglect and overt criticism, it is now increasingly viewed as a useful technology. It is incorporated in commercial process controllers as well as a number of application-specific controllers for consumer products.

In fact, fuzzy logic is becoming commonplace in Japanese consumer products. Examples include electronic ''jiggle'' removal to provide a steadier picture in a hand-held VHS-C camcorder. The camcorder's charge coupled device (CCD) image detector captures and stores video frames 20 percent larger than the view seen in the viewfinder. The camcorder compares successive frames by identifying correlations between frames. The camcorder then crops each frame as required to minimize jiggle-induced variations from frame to frame, and thus produces a stabilized frame of the correct size.

In other applications, a Mitsubishi room air conditioner uses fuzzy logic and is claimed to cut power consumption by 20 percent. A Matsushita fuzzy logic vacuum cleaner varies motor power from 100 to 260 watts in response to the type of floor and amount of dirt. A row of four lights indicate the rate at which dirt is being removed. The user of a Hitachi or Matsushita fuzzy washing machine need only press a single button to start the process. Fuzzy logic analyzes the output of infrared sensors to determine the size of the load (to set the water level), whether clothes are light or dark (to set water temperature), whether clothes are delicate or not (to set agitator speed), and whether dirt is still present (to set the length of the wash cycle). Fuzzy logic thus automatically selects one of the 600 water level and washing cycle combinations available.

Automotive applications include controllers for antilock brakes, automatic transmission shift scheduling, and speed control. General Motors joined Japanese brands in these applications.

Omron Electronics, a supplier of **programmable logic controllers (PLCs)** and single-loop controllers, is a major proponent of fuzzy logic for process controllers. Some of their PLC and single-loop controller products incorporate fuzzy logic, which is claimed to be particularly useful for the following types of applications:

1. Processes with changing characteristics or environments for which adaptive control is required.
2. Nonlinear processes such as tensioning and position control.
3. Difficult-to-control systems typically requiring human judgment.
4. Processes with conflicting constraints or multiple inputs.
5. Processes with large input deviations or for which these deviations are measured with low resolution (or high uncertainty).

FIGURE 4–2
Crisp Comfort Membership Function

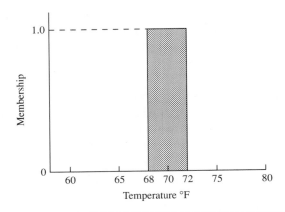

Similar claims are made for neural net-based control. The theme common to the two approaches is better accommodation of system behavior such as nonlinearity that in many cases is not handled well by conventional control techniques and the simplifying assumptions that they incorporate.

Fuzzy logic and expert systems are conceptually related in that both are rule-based. The difference between the technologies lies in the way that uncertainty is handled. Expert systems require precise, or **crisp** quantitative rules. Fuzzy logic users, in contrast, may state qualitative, or **fuzzy rules.** The fuzzy logic system then processes these qualitative relationships and ultimately generates a specific quantitative output.

The net effect is that relationships that are known, but not exactly, are more easily handled with fuzzy logic. Expert systems designers may need to add many (crisp) rules to accommodate uncertainty in data or action. The relationships expressed by fuzzy rules indicate objectives, but with an accommodation of uncertainty. From a system development perspective, this may enable fewer rules to accomplish the same goal.

The qualitative nature of fuzzy logic may also enable goals to be more easily expressed in this way. For example, a control objec-

FIGURE 4–3
Fuzzy Comfort Membership Function

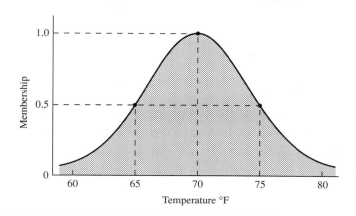

tive of achieving and maintaining a comfortable room temperature is clear and unambiguous in a qualitative sense. People know whether they are comfortable, too cold, or too hot, and thus whether the temperature should be raised or lowered. As the temperature deviates further from a comfortable norm, people sense that a greater control action is required. This fuzzy control specification is more intuitive than a more common crisp statement such as below 68°F is too cold, and above 72° is too hot—period.

In traditional logic, control actions would be based on the system's current state relative to, for example, temperature setpoints of 68° and 72°. In fuzzy logic, control objectives are defined by smoothly varying membership functions. Control actions are based on a system's current state relative to that membership function. Figures 4–2 and 4–3 illustrate the distinction.

Figure 4–2 shows a crisp membership function—in this case, for comfort. Temperature is judged to be comfortable only if it is between 68° and 72°. This is represented by a membership function equal to 1 between 68 and 72, as shown on the vertical axis, and 0 elsewhere.

Figure 4–3 reflects the more realistic fuzzy scenario. The further the temperature deviates from 70°, the less comfort is perceived.

At 65° and 75°, the membership value is 0.5 signifying less comfort than that perceived at 70°. This does not imply absolute discomfort, but only less comfort.

This may appear to be merely an exercise in semantics. But in fact, this approach is helpful in, for example, distilling many individuals' perception of comfort into a single satisfactory-to-all control strategy. In general, the appeal of fuzzy logic is that it enables simpler specification of control objectives and thus faster system implementation.

In this case, fuzzy logic provides an easy way to state the control engineer's objective to keep the process variable, temperature, centered within a range and to implement control actions that diminish in magnitude as the setpoint is approached, so that overshooting is avoided.

Fuzzy logic has found application primarily as a control technology. In the applications for which it is best suited, the benefits include easier programming because of a more natural expression of control objectives. This, in turn, enables more rapid prototyping, system development, and debugging. In the simple temperature control case, the objective could be stated as ''Keep the temperature close to 70°.'' The fuzzy logic controller would implement control actions of progressively smaller magnitude as the system approaches the setpoint. This contrasts with the multiple rules required in an expert system approach or the equations required in a traditional approach to define the smaller control actions to be taken as the setpoint is approached.

Rules in fuzzy logic systems perform differently than in expert systems. Expert systems consist of a hierarchical structure of rules called a **decision tree.** Users solve problems by answering questions posed by the expert system that define the current situation and the problem to be solved. Every answer provided by the user enables the expert system to focus on increasingly limited domains. This may be compared to climbing a tree, making a decision to leave the trunk on one of many major branches, and continuing a decision-making progression of choosing ever-smaller branches. This progression leads the user to a single rule that covers the situation at hand. Action or advice is provided on the basis of that single rule.

In contrast, fuzzy logic systems are characterized by the ability to fire, or act upon, multiple rules at once. Several rules may be

FIGURE 4–4
Fuzzy Cold and Hot Membership Functions

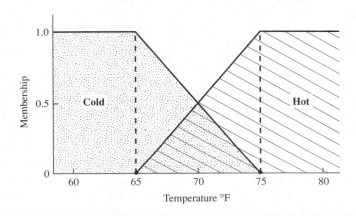

true, to a degree, for the situation at hand. The fuzzy logic system fires all of the rules that apply, each to the degree that it applies. In this way, qualitative (fuzzy) objectives are translated into specific (crisp) actions.

This can be illustrated by expanding the previous temperature control application to include control of both heating and cooling, for commercial buildings rather than private residences. The control problem becomes more complex, but the solution more optimal in terms of comfort and efficiency.

Home heating and air conditioning employ *bang-bang* control. Either the furnace is on, if heating is required, or it is not. Either the air conditioning is on, if cooling is required, or it is not. Unlike these discrete control action choices of heat, cool, or do nothing for a residential application, the possible control actions for the commercial application would be continuous from high heat rate, through no action to high cooling rate.

Defining appropriate control objectives from a fuzzy logic perspective might involve two membership functions for temperature, as shown in Figure 4–4.

The cold membership functions show perception of coolness. Figure 4–4 shows that below 65° everyone is cold (membership = 1.0), and above 65°, the declining membership value shows that, collectively, people feel less cold. The hot membership function illustrates a similar relationship regarding perception of hotness. The controller uses these functions to make a decision on whether to call for heat or cooling, and to what degree.

At a given temperature, the fuzzy controller analyzes the degree of membership in both functions to determine whether the control rules associated with one function, the other, or both should be fired. Figure 4–4 shows that in the region between 65° and 75°, there is membership in both functions. Simultaneous membership in both functions requires that the rules of both functions be explored by the controller to establish a control action appropriate for a given current temperature. Only one control action is taken, but this action is based, in this case, on the simultaneous firing of several rules.

The advantage of this approach is that each of the rules is broader in scope, so that fewer rules are needed than in an expert system, in which each of the larger number of rules tends to cover a narrower range of conditions. The disadvantage of this approach is that fuzzy logic rules can require processing periods 30 to 40 times that of expert systems. This is because fuzzy logic controllers must perform an operation called **defuzzification** to translate membership in one or more fuzzy sets into a crisp, specific control action. After identifying the appropriate rules in a given circumstance, the controller must then use one of several mathematical approaches to determine the weights, or relative impact, of the outputs or each rule multiplied by its associated weight and all weights summed to provide the crisp output. This can involve intense computation since multiple membership functions may result in a large number of rules to be tested and calculations to be performed.

For this reason, although fuzzy logic can be implemented in software alone, many real-time control applications use special hardware to process rules more rapidly. As an example, the previously described Mitsubishi fuzzy logic room air conditioner uses a Togai Infralogic FC 110 fuzzy logic chip.

Engineers may consider as an alternative a hybrid system in which time-sensitive aspects of control are handled by high-speed nonfuzzy hardware and software, and the fuzzy logic portion of the controller interacts with the high-speed portion in a way that does not affect the control system's response time.

The stability of fuzzy-logic–controlled systems cannot be proved mathematically. This is also true of other approaches such as some neural nets, and in most cases is of no particular practical consequence. However, in safety-related and other critical applications, this is a problem. In these cases, a necessary implementation step is to perform exhaustive testing to verify system performance under all possible circumstances.

An example is provided by a fully automated subway train with fuzzy logic control that has been in operation in Japan since 1988. The fuzzy logic controller is reported to make almost 70 percent fewer errors in braking and acceleration than humans and to provide an energy savings of 10 percent. But implementation was a challenge. The safety implications are considerable in this control application. The inability to mathematically prove control stability—that is, to prove that there is no possible scenario under which the control system would respond inappropriately—required drastic measures. Observers were on board for tests consisting of 300,000 runs! At this point, it was agreed that safety had been sufficiently demonstrated, and the system was commissioned.

A final application issue for both fuzzy logic and expert systems may be belaboring the obvious, but that is the need to know appropriate rules on which to base recommendations or actions. In many cases, rules are readily available. In others, rules are ambiguous or incomplete. In these cases, it may be appropriate to call upon other technologies to discover underlying relationships on which strategies and rules can be based. This need to discover relationships is addressed by another knowledge-based technology called neural networks.

4.6 NEURAL NETWORKS

Neural networks are a completely different approach to computing. They are also known as **artificial neural nets (ANNs)** and simply neural nets. They are lumped together with expert systems and fuzzy logic in the broader category of knowledge-based systems because they share the common goal of enabling computers to capture and apply knowledge, not just manipulate numbers and letters.

Neural nets have a number of interesting properties. They can improve their own performance; they can adapt; they can discover relationships in data. This means that a computer-based system can now learn by itself to design better products and perform manufacturing tasks more effectively.

As an example, a common design problem for products made according to a recipe is to find ways to vary the recipe, or proportions of various ingredients, to provide new product properties. If a neural net is provided recipe data, along with the corresponding properties data, it can discover the relationship between the two. In this case, that relationship is a product formulation model that can be used to design new or better products. The user need not know how changes in recipe affect changes in properties. The user need only provide data that captures that relationship.

This discovery capability provides an interesting alternative to dependence on rules-based knowledge. Now, corporate databases can be mined to discover knowledge, not just numbers. This knowledge includes data such as marketing information, product designs, manufacturing records, and financial results. Each of these domains is interesting in its own right. But an opportunity exists to discover interesting relationships among, say, how a product was designed, how it was manufactured, how it performed functionally, and how it performed as a source of revenue.

Neural nets differ from expert systems and conventional computer technologies in that they are not procedural in nature. The essence of conventional programming is to specify procedures to be applied to data. The programmer explicitly defines the operations and the bounds on these operations to accomplish the desired computational goal. (The challenge, of course, is to develop procedures that fully meet goals and do not have unintended consequences, also known as *bugs*.) Computer languages enable programmers to manage the flow of data, define procedures, and manage the interaction among these procedures. Each program has a start, a logical progression, and an end.

Neural nets provide an interesting contrast. They are not programmed as such but are taught by example. Neural nets learn to solve a problem by being shown examples of situations and associated solutions. The net learns a relationship that is valid between each of the sample situations and the associated solutions. Upon finding this relationship, the net is able to generalize to provide an appropriate solution to a new situation. This is a powerful capability that addresses one of the basic goals of AI: to expand computer capabilities by enabling the computer to learn. By discovering relationships in data, neural nets implement machine learning.

The inspiration for neural networks came from studies on the structure and function of brains and the mechanisms of learning and problem solving. Neural nets take their name from the system found in the brain of a large number of simple processors called **neurons** which are interconnected by a network that transmits signals between them. Computer-based neural nets are, to a degree, a simulation of the genuine biological article. However, the analogy can be drawn only so far. For reasons that are not fully understood, the performance of computer-based neural nets falls far short of the brains they simulate.

But engineers tend to view neural nets from a different perspective, bypassing biological analogies and focusing instead on the unique properties and problem-solving capabilities of neural nets. Neural nets are inherently adaptive, nonlinear, and pattern recognizing in nature.

The adaptive capability is a result of the self-discovery capability. If the net can automatically discover a complex relationship between, say, current process parameter values and ensuing process results, the net can as easily discover the new relationships caused by evolving equipment characteristics or changes in environment or raw material properties.

Neural nets are inherently nonlinear. This enables them to model complex relationships without resorting to the simplifying assumptions that are commonly used with linear approaches, such as approximating a nonlinear relationship with a linear but limited range relationship or by representing that relationship by piecewise linearization.

The pattern recognition nature of neural nets provides greater capabilities and speed in a number of application areas. Patterns to be recognized may include what are commonly considered to be images, such as radar, CAT scan, visual images, or ultrasonic test signatures.

By considering a pattern in its entirety, rather than detail by detail, this approach is not only faster but often better able to recognize underlying relationships without being distracted by detail differences. In inspection, this characteristic enables neural net-based inspection systems to be less likely to erroneously reject good parts. In other applications, this characteristic makes neural nets unusually tolerant of noisy inputs, because the net recognizes the pattern underlying the superimposed noise.

These pattern recognition applications are examples of one broad area of neural net functionality—classification. The classification task is to determine which of several categories a new circumstance most closely matches. A neural net can determine which prior image a new image most closely matches and thus validate signatures on checks. A neural net can also determine which prior process control scenario the current situation most closely resembles and thus identify appropriate setpoints and monitoring procedures.

The other broad neural net capability is that of estimation. The estimation task is to calculate an answer based on a functional relationship. In this case, a neural net learns a relationship that serves as a model of a problem to be solved. That net is then consulted, or requested to provide an estimated response to a given input. One example of estimation is to use a neural net-based formulation model to determine the effect of a change in product formulation, or recipe, on the properties of that product. Another example in the area of process control is to estimate a future system state, given current and past system states.

A neural net does these things by learning a relationship between numerical inputs—for example, product recipes—and numerical outputs—for example, product properties. A neural net is enabled to solve problems by discovering that relationship or model and then using that relationship to transform new inputs—for example, a new recipe—into new outputs—for example, product properties.

This is accomplished with either hardware or, more commonly, software that simulates simple processors, called neurons, interconnected in one of many possible architectures. Figure 4–5 illustrates a typical arrangement.

At the bottom of the figure, inputs are presented to the net. These numerical values may represent, for example, relative proportions of four product ingredients. The outputs, as shown at the top of the diagram, might be two properties of the resulting product, perhaps tensile strength and ductility.

Each simple processor, represented by a circle, adds together signals received from inputs or other processors, according to ar-

FIGURE 4–5
A Neural Net

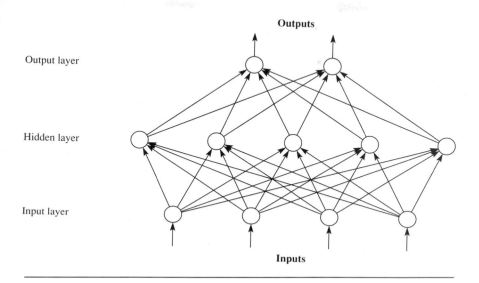

Outputs

Output layer

Hidden layer

Input layer

Inputs

rangements shown by the arrows. The incoming signals are added, modified by a transfer function, and transmitted to outputs or other processors by the outgoing arrows. Each arrow acts as a multiplier of the signal it transmits, according to the weight assigned to that individual arrow. A neural net is able to provide the appropriate transformation of inputs into outputs by finding the proper weight for each arrow, so that the series of input numbers presented to the net are correctly transformed into output numbers by the additions at neurons and multiplications along the arrows.

The trick is to find the values of these weights that will accomplish the desired transformation. The neural net is said to be trained when it is able to provide this transformation. This is done by a computationally intensive process in which perhaps hundreds or even millions of values are tried for each of the weights to arrive at a combination of weights that provides an acceptably low difference between the actual and the desired transformation. The assessment of this difference, or error, is made using sample data for which the correct answer is known.

An initial set of weights is chosen at random. Input data is taken for the example problems from which the net is to be taught. The neural net performs summations at neurons and multiplications along arrows and generates an output. The net tests to see how close the predicted answer, or output, is to the correct answer of the example problems. Of course, in the first try, the error is high. A second set of weights is selected and the net recalculates the output. Again, the error is determined. The net repeatedly tries different sets of weights, calculates the error, and determines if the selection of weights is getting better or worse as determined by the error getting smaller or larger. The net varies weights in the direction that is found to make the error decrease, finally settling on weights that provide acceptably accurate training.

Typically, hundreds to thousands of iterations are required to bring the error to an acceptably low level. This ordinarily requires from minutes up to hours, depending on the computer, for realistically large problems and without the use of special neural net hardware. Once the net has been trained, it is available to solve problems by estimation or classification. This consulting of the net is virtually instantaneous, consisting of a single series of multiplications and summations, not the many required for training. The following discussion thus applies only to initial system training, not to reuse of the trained net.

Several factors affect training time. One is the speed of the computer. A second factor is that of neural net training parameters. Values must be chosen for these parameters, and these choices affect training time. The challenge is that the best value for these parameters depends on the problem at hand. If these parameters are not automatically selected by the neural net and the developer does not have the experience to set them correctly, training times can be extended. In these cases, millions of interations and overnight training times can be required. A third factor in training time is that of neural net algorithms. The possibility of excessive training time has provided the incentive to develop new, more efficient neural net approaches. Algorithms such as the Functional Link Net, described below, and others have been shown to divide training requirements by factors of over 1,000. In addition, some problems can be decomposed, or divided into several simpler problems, to further reduce training time and improve accuracy. Research in these and other areas continues in pursuit of faster training.

Many variations are possible from the neural net architecture of Figure 4–5. More nodes may be required per layer to accommodate more inputs or outputs. More hidden layers may be employed to enable representation of more complex relationships such as nonlinearity or large dimensionality. An architecture that is too simple will not capture the full complexity of a problem. That is, the neural net-based model will not accurately predict the response of the real system.

Unfortunately, a too-complex architecture (with too many neurons) also leads to modeling errors. In this case, the neural net portrays the relationship as more complex than it actually is. The effect is one of adding nonlinearity or, in essence, modeling noise.

As in most technical areas, there is no substitute for technically competent judgment in identifying applications and setting up solutions. Judgment is also required to provide a reality check on the answer. This enables the developer to verify that the proper net architecture has been selected and that other setup and solution steps have been properly done.

Expert guidance is helpful in selecting an architecture. Such guidance includes rules of thumb derived from both theoretical considerations and application experience. However, in the absence of this guidance, a developer may identify the best architecture by a process of trial and error. The goal is to select the minimum number of neurons and layers that can fully represent the underlying relationship to be modeled. This is tested by comparing the results of the neural net-based model with known answers over the anticipated range of use of the model.

Validation should be performed by comparing neural net predictions to data obtained by experimentation or application. These data may be difficult to obtain, and so it is tempting to use a statistical model to generate validation. The risk in this approach is that there is generally no basis on which to assume that a statistical model is more accurate than a neural net model, and thus no conclusion can be drawn from the comparison.

If the model validation indicates that the model is too complex, adding nonlinearity or noise to a functional relationship, the neural net must be simplified by removing neurons and perhaps hidden layers. If validation indicates that the model is too simple, not capturing the full detail of the modelled relationship, the neural net must be expanded.

One common way to do this is to add more layers and with these layers more neurons. However, this creates a problem in that the time to train the neural net climbs radically as the net architecture expands. This is due to the combinatorial explosion that occurs when all combinations of the now-larger number of interconnections must be explored.

An alternative approach is to add inputs that enable representation of nonlinearities or interactions between variables. This can be done by making the net wider, with more inputs but no more layers. In this way, complex relationships are represented, but training times are minimally increased. The new inputs must be selected to enable representation of complexity without adding new complexity or otherwise coloring the relationship. This may be accomplished by use of the Functional Link Net architecture, patented by AI WARE. This architecture enables users to choose *functional enhancements,* creating additional inputs based solely on the original inputs.

If the user suspects that the neural net output might be a function of the square of an input, the user can create an input that consists of one of the original inputs squared. Similarly, the user might suspect that the neural net output is a function of the interaction between two inputs and thus create an input that is the product of the two existing inputs. Finally, the user may create a new input that consists of an original input multiplied by a (nonlinear) trigonometric function. This enables the network to represent a nonlinear relationship in a single hidden-layer net without the compromise in performance that results from the alternative of a larger multilayer network that is usually required to represent nonlinear relationships. In each case, the user may choose functional expansions based on known or suspected underlying relationships. Alternatively, the user may try several functional enhancements and select the one providing the fastest training.

As with expert systems, neural nets perform better if the user can give the system hints on the nature of the problem. In the case of expert systems, hints are incorporated in decision-tree structures that reflect the structure of knowledge being captured. With neural nets, hints are acted upon with inputs expanded to reflect relationships such as square law or other nonlinearities suspected to be part of the relationship being modeled.

Neural net performance may be improved in some cases by recognizing that what appears to be one problem is actually several subproblems that share common inputs and outputs but differ in the exact relationship between input and output. This might be viewed as different best ways to solve a problem depending on the region in data space in which the inputs are located. For example, there may be several good process windows, each providing equally good results, but each best suited to a particular range of controlled and uncontrolled variable process values.

This opportunity is pursued by first identifying clusters of data that represent some type of similarity between some data points but not others. A relationship may involve many different types of similarity, each represented by a data cluster. One cluster might represent surface finish on parts milled at X feet per minute. Other clusters might represent results for milling speeds of 1.2, 1.4, and 1.6X.

Neural nets find clusters of data through a process called **unsupervised learning.** This term refers to the fact that the net is looking for relationships in data without any guidance, or supervision, regarding possible relationships. This contrasts with **supervised learning,** in which the net is asked to find a known (or suspected) relationship between some variables presented as inputs and others presented as outputs.

Supervised and unsupervised learning generally requires different neural net architectures. Not only must the user select the two architectures, the results of the unsupervised learning process must be explicitly transferred to the supervised learning environment. A benefit of the Functional Link Net approach is that it provides a common architecture for both functions, and so unsupervised and supervised learning are accomplished within one integrated environment.

Neural nets may be viewed as modeling technique, in that they develop a mathematical representation of a particular system that can be used to predict the performance of that system. Neural nets thus encounter the same issues as other modeling techniques such as first principles and statistics.

4.7 MODELING ISSUES AND APPROACHES

When making decisions, the best possible outcome is sought by calling upon an understanding of the problem at hand. Hopefully this understanding includes knowledge of relevant factors and the way that these factors are related. If this understanding of a situation is sufficient to enable prediction of the result of a particular action, a decision support model exists.

The model may be implicit. The relationship between seasons of the year and weather patterns is well understood. Without rules or equations, one can anticipate when snow tires will be needed. But many decisions are more difficult because a decision support model is not obvious. How should coating products be reformulated to eliminate volatile organic compounds, or friction materials modified to eliminate asbestos? How should a product be designed to maximize market share? How should a process be run to maximize yields, minimize maintenance, or both?

Many decisions have significant performance, cost, market share, and even regulatory impacts, and for these reasons, it can be well worth the bother of developing a good decision support model. There are several approaches to model development. Traditional approaches of first principles and statistical analysis are widely used. Neural networks are gaining acceptance as a credible alternative.

In high school physics classes, one encounters equations describing the relationship between force, mass, and acceleration. Chemistry rules and equations describe the ways that chemicals combine and react. These so-called first principles are fundamental, universally applicable rules of behavior. A model based on first principles inspires confidence because it is based on universal truths.

The use of first principles has two limitations. First, many real-world applications are simply too complex to model with first principles. The dimensionality of the problem may be high. The exact nature of the interactions among variables may be unknown. In many chemical formulation appplications, for example, ingredients interact both chemically and physically. Experts know that these interactions occur, but in real-sized problems they often cannot disentangle the many interactions sufficiently to explicitly model them.

A second limitation in the use of first principles is that it is often hard to accurately characterize secondary effects. In process monitoring and control applications, for example, secondary effects such as friction and other equipment characteristics may change over time due to the effects of age, degradation since maintenance was performed, or perhaps undocumented changes in hardware configuration. In the model, these effects may be ignored or may be assumed to be a constant, nonzero value. These may or may not be reasonable assumptions.

Often there is no way to know for sure because no procedure exists to identify and correct models that have become inaccurate over time. This can be a significant issue in equipment with long service lives. As an example, industrial boilers may remain in use for 50 years. During this time, heat transfer characteristics change slowly due to the accumulation of combustion by-products. Combustion process control strategies may be based on heat balance models that do not reflect these changes and thus no longer provide optimal efficiency and emission levels.

These problems have provided the rationale for the most common modeling alternative—statistics. The goal of statistics (and of neural nets) is to model relationships too complex or opaque for first principles. The approach is to find observable and repeatable relationships between parameters assumed to represent *cause* and those presumed to represent *effect*. There is no basis on first principles, only on observed relationships in data.

One common application, **statistical process control (SPC),** involves regular measurements of process parameters. Statistical measures of these parameters, such as the mean value and the variations about that mean, are calculated and plotted. Observation of trends represented by these plotted numbers provides process operators early warning of problems before process parameters exceed acceptable limits.

In the same vein, **statistical quality control (SQC)** involves statistical analysis of quality parameters. Analysis of trends provides early warning of what will become quality problems if timely corrective action is not taken.

Statistical models may be based on known interactions. For example, an understanding of relevant first principles may help select the form of the statistics equation. But this level of fundamental understanding may not be available, and so the statistical model will instead be based only on relationships observed in data.

In this case, the same issues regarding data completeness and quality apply to statistical methods as to neural nets. Since the model is based solely on data, the model's accuracy is limited by the accuracy and completeness of data. Data must cover the full range of parameter values and must be complete and of good quality, or steps must be taken to correct these faults. Such steps are often required because real-world data tend to have gaps, transcription errors, and noise, or random errors. Noise may be due to faulty process sensors or experimental procedures that fail to control the environment sufficiently to provide repeatable results.

Development of a statistical model may be viewed as curve fitting. An equation is sought that will generate a line passing acceptably close to known data points. This process involves selection of the correct form of equation, followed by identification of the correct values of the constants in the equation. For example, a simple linear relationship may be expressed as $y = Ax + B$. Values of A (slope) and B (y-axis intersection point) would be found to generate a line that best fits known data.

Conceptually, at least, this process can be applied to high-dimensionality problems by finding a surface that fits known points in n-dimensional space. With simple relationships and noise-free data, the equation form may be readily apparent. In more complex cases, it may not. Expertise in both statistics and the problem at hand is required to accurately judge the appropriateness of the equation form. Judgment may be made on the basis of known underlying relationships or on the basis of analysis of data. If the correct form is not chosen, the resulting model will be flawed. The need for judgment and the impact of poor choices are the statistics counterpart to the similar issues regarding selection of the proper neural net architecture.

Reports on the relative merits of statistics and neural nets by users of both indicate strengths in statistical methods, including widespread familiarity among users and a great variety of supporting products. Application-specific techniques such as statistical process control also provide clear benefits.

Users also report benefits of neural nets over statistics. Statistical methods are subject to error in instances with sparse data, particularly when those data do not include data points representing all combinations of the extreme values of each variable. This situation is not uncommon when available resources preclude a formal **design of experiments (DOE)** program and collection of the entire data set specified by the DOE program.

Statistical methods are also subject to error in cases with intercorrelation between independent variables. Statistical methods encounter difficulty because it is quite difficult to separate the individual influences of each variable. A practical, if not particularly accurate, approach in these cases is to treat the problem as a large number of independent single-variable problems rather than one multivariable problem. This simplifying assumption can lead to significant modeling errors in some cases.

Neural nets, in contrast, can handle multiple interacting variables as such without artifically simplifying the problem. Mixture problems, such as determination of product formulation, are an example of this type of problem. The proportions of ingredients in a mixture are clearly independent. Yet, the effect of each ingredient can depend on the level of each of the other ingredients. This dependency is caused by chemical and physical interactions between ingredients. Users report that due to the difficulties with statistical methods in accommodating these interactions, neural net models are more accurate than statistical models for mixture problems.

First principles, statistics, and neural nets each have their strengths in modeling applications. Users report the following:

- Models based on first principles are preferred when feasible.
- Statistics is a more established technology than neural networks, and therefore more comfortable to many.
- Statistics provides good results in lower-order relationships involving only independent variables.
- Neural networks provide better results in multidimensional and time-varying situations and when variables exhibit interdependencies.
- Neural networks are more effective than statistics with data not meeting rigorous design of experiment criteria. In particular, neural nets have been shown to provide more accurate models when data are not available for all combinations of extreme values of all variables.

These findings should be considered when identifying potential KBS applications. Another point to be considered is the ways that knowledge-based technologies can be combined to accentuate the positive and minimize the negatives of each.

4.8 INTEGRATED SYSTEMS

Expert systems and neural nets are dissimilar yet complementary. Assembling the rules for an expert system is tedious at best, assuming that a cooperative and concise expert is available. But the use of explicit rules enables clear-cut explanations to be provided on the logic behind a decision. Neural net systems, on the other hand, can be implemented rapidly and maintained easily. Although they do not provide an explicit explanation capability, their adaptive and self-improvement capabilities make them uniquely suited to many applications in design and manufacturing.

The capabilities of expert systems and neural nets may be combined in several ways. A common strategy is to embed one or more neural networks within an expert system. The expert system functions as the interface to the user. The expert system also contains rules to define and divide problems into subproblems, prepare data for neural net analysis, configure neural nets, and analyze their outputs. The nets' output, a series of numbers, can then be translated back into text-based recommendations, if required, by the expert system, which then presents text and perhaps graphics output to the user. This arrangement combines the information presentation strengths of expert systems and the analytical capabilities of neural nets.

Fuzzy logic and neural nets are very different in nature. Fuzzy logic explicitly expresses knowledge with IF-THEN rules, and is suited to higher-level reasoning. Neural nets capture knowledge in the form of relationships within (numerical) data.

But there are striking similarities between the technologies. Neural nets learn relationships that often look like fuzzy logic membership functions. These functions describe relationships between two variables, on the basis of which a control action or decision will be based. A neural net-enabled opportunity would be to discover the nature of the relationship—in particular, to discover the shape of the membership function curve. In this sense, a neural net acts as the front end of a fuzzy logic-based control system. Control actions are based on fuzzy rules with neural net-discovered membership functions. This approach is reported to be actively pursued by Matsushita and other major Japanese companies, as well as the Japanese Ministry of International Trade and Industry(MITI). Reported benefits are reduction in the time required to design systems with one (undocumented) example given of a reduction from 30 hours to 30 minutes.

An alternative fuzzy logic/neural network collaboration would be to use neural nets to act upon the results of a fuzzy logic analysis. The action of a neural net, summing at nodes and multiplying along links, is similar in some ways to the fuzzy logic process of defuzzification. This process translates the partial outputs of several fuzzy rules into a specific control output or decision by taking a weighted sum. A neural net could thus perform the time-consuming process of defuzzification.

This approach is reported to be receiving less attention in Japan. However, an American company, AI WARE Incorporated, offers a line of software products that expand upon this approach. These products are targeted to applications such as product design and process optimization, and are not sold as fuzzy logic tools. Users specify preferences in product properties or process optimization objectives through **desirability functions** that are in effect fuzzy logic membership functions. An integrated neural net is used to develop a design or process model. A built-in data preprocessor builds on the capabilities of statistics to measure and improve data quality. The design or manufacturing model is used in conjunction with the desirability (membership) functions to drive an integrated optimizer. The optimizer output representing, for example, optimal product designs or process settings meets multiple ranked objectives, while observing constraints in cost, design, and production. This approach, as will be described in upcoming chapters, combines the benefits of neural nets and fuzzy logic and provides capabilities not practical with traditional approaches or with neural nets or fuzzy logic alone.

This chapter has outlined a wide range of problem-solving capabilities based on three distinct technologies. This information will help the reader match technologies and opportunities. When combining technologies, as when building any system, the analysis should be made of needs first and solution technique second. The choice of which technologies to use, knowledge-based and otherwise, must be made on the basis of the problems to be solved and the knowledge resources available. System developers and their management want to know whether given technologies best meet their needs, if their expectations are realistic, and how others have fared in similar pursuits.

The following chapters provide applications examples for each of the technologies described. The focus is on solutions rather than technologies, and for this reason the chapters are grouped by application area, beginning with the design of discrete parts. Applications examples for each technology help the reader compare technical capabilities in the context of specific applications needs and constraints. The history of knowledge-based systems in each application area is followed by guidelines on identifying and implementing knowledge-based systems. This information will help the reader understand how and why new applications should be pursued.

4.9 REFERENCES

1. AI WARE Incorporated company literature. Cleveland, OH, 1993.
2. Arnold, William R., and John S. Bowie. *Artificial Intelligence: A Personal, Commonsense Journey.* Englewood Cliffs, NJ: Prentice Hall Inc., 1986.
3. Babb, Michael. "Fast Computers Open the Way for Advanced Controls." *Control Engineering,* January 1991, pp. 45–51.
4. Bartos, Frank J. "Fuzzy Logic Widens Its Appeal to Industrial Controls." *Control Engineering,* June 1993, pp. 64–67.
5. Bylinsky, Gene. "Computers That Learn by Doing." *Fortune,* September 6, 1993, pp. 96–102.
6. Caudill, Maureen. "Expert Networks." *Byte,* October 1991, pp. 108–16.
7. Caudill, Maureen. "Neural Networks Primer, Part I." *AI Expert,* December 1987, pp. 46–52.
8. Caudill, Maureen. "Neural Networks Primer, Part II." *AI Expert,* February 1988, pp. 55–61.
9. Dreyfus, Hubert L. *What Computers Still Can't Do: A Critique of Artifical Reason.* Cambridge, MA: The MIT Press, 1992.

10. Ferrada, Juan J., and John M. Holmes. "Developing Expert Systems." *Chemical Engineering Progress,* April 1990, pp. 34–41.
11. "Fuzzy Logic: A 21st Century Technology." Distributed by Omron Electronics, Inc., 1991.
12. Gill, Tyson, and Joel Shutt. "Optimizing Product Formulations Using Neural Networks." *Scientific Computing & Automation,* September 1992, pp. 19–26.
13. Grant, Eugene, and Richard S. Leavenworth. *Statistical Quality Control.* 4th ed. New York: McGraw-Hill Book Company, 1972.
14. Hamburg, Morris. *Statistical Analysis for Decision Making.* New York: Harcourt, Brace & World, Inc., 1970.
15. Hammerstrom, Dan. "Working with Neural Networks." *IEEE Spectrum,* July 1993, pp. 46–53.
16. Hillman, David V. "Integrating Neural Nets and Expert Systems." *AI Expert,* June 1990, pp. 54–59.
17. Johnson, R. Colin. "Japan Clear on Fuzzy–Neural Link." *Electronic Engineering Times,* August 20, 1990, p. 1.
18. Keyes, Jessica. "Where's the 'Expert' in Expert Systems?" *AI Expert,* March 1990, pp. 61–64.
19. Keyes, Jessica. "Why Expert Systems Fail." *AI Expert,* November 1989, pp. 50–53.
20. Klassen, Myungsook. "Incremental Learning for Recognizing Handwritten Characters Using Neural Networks." Technical Report TR89–113. Cleveland, OH: Center for Automation and Intelligent Systems Research, Case Western Reserve University.
21. Lipkin, Richard. "A Discipline Seeks to Grasp the Fleeting Silicon Logician." *Insight,* February 15, 1988, pp. 13–17.
22. Mehta, Chetan. "The Business Link: Integrating Process, Economic and Contractual Data Will Help Optimize Operations for Maximum Profits." *Chemical Engineering,* May 1992, pp. 96–100.
23. Morrow, Mike. "Genetic Algorithms: A New Class of Searching Algorithms." *Dr. Dobb's Journal,* April 1991, pp. 26–32.
24. Neural Ware product literature, Pittsburgh, Pennsylvania.
25. Pao, Yoh-Han. *Adaptive Pattern Recognition and Neural Networks.* Reading, MA: Addison-Wesley, 1989.
26. Pugh, G. Allen. "A Comparison of Neural Networks to SPC Charts." *Computers ind Engng* 21, nos. 1–4.
27. Rauch-Hindin, Wendy B. *A Guide to Commerical Artificial Intelligence—Fundamentals and Real-World Applications.* Englewood Cliffs, NJ: Prentice Hall, 1988.

28. Roberts, Marcus. "Twelve Neural Network Cliches." *AI Expert,* August 1988, pp. 40–46.
29. Ross, Randy. "An Expert System That Failed." *High Technology Business,* May 1988, p. 25.
30. Schmuller, Joseph. "Three Faces of Fuzziness: Theory, Practice, and Applications," *PC AI,* March–April 1993, pp. 14–15, 27–29.
31. Sobajic, Dejan J., and Yoh-Han Pao. "Neurocomputing in Power Systems," *INNS above Threshold Newsletter,* March 1993, pp. 10–13.
32. Stock, Michael. *AI in Process Control.* New York: McGraw-Hill Book Company, 1989.
33. Studt, Tim. "Neural Networks: Computer Toolbox for the '90s." *R & D,* September 1991, pp. 36–42.
34. VerDuin, William H. "Neural Nets: Software That Learns by Example." *Computer-Aided Engineering,* January 1990, pp. 62–66.

Chapter Five

Knowledge-Based Discrete Part Design:
Incorporating Design and Fabrication Expertise

5.1 INTRODUCTION

The previous chapter on neural nets, fuzzy logic, and expert systems described the problem-solving capabilities of each. This chapter is the first of four that describe applications of these technologies. The focus of this chapter is on the ways that knowledge-based technologies support the design of discrete parts. Chapter 6 also covers knowledge-based product design but for products that are produced by batch or continuous processes.

Discrete parts are products that are manufactured as individual, discrete parts, and so their production is measured in counts of individual items. Discrete parts such as fasteners may be sold as such or may be incorporated in complex assemblies such as computers or automobiles. This contrasts with products made by batch or continuous processes, whose production is measured by volume or weight. Examples of this class of products include metals, plastics, and chemicals.

Traditionally, discrete part design was considered to be primarily the manipulation of geometries. A revised part might be only a longer or shorter version of the previous part, or it might involve a completely new design. But the essence of the task was to conceive and draw a new configuration. At this point, material specifications might be reviewed and occasionally changed. Processing would generally not be addressed because design and manufacturing were seen as unrelated tasks.

The new design would be validated through testing. Modifications would be made to design or materials if required. The design was then presented as an accomplished fact to the manufacturing staff.

This simple scheme is falling from favor. Manufacturers are viewing the design task in a broader context, as they seek to meet new requirements in product quality, features, and cost. They do this by pursuing new materials and processing opportunities that better meet product requirements while minimizing costs.

The broader scope of design decisions and the need to shorten time to market have led to the coupling of design and manufacturing in an integrated process known as **concurrent engineering,** and more recently as **integrated product/process design (IPPD).** This new approach is contrasted with the traditional approach in Figure 5–1.

The traditional product design approach is based on geometry as the primary design issue. Changes in material and processing are considered sequentially or not at all. The need to revise material may be identified but perhaps only as a result of a failed test of a prototype part. By considering geometry, material, and then process sequentially, the entire design process may require several redesigns or changes in material and subsequent retesting. Opportunities to make best use of new materials may be missed by not considering material and geometry at the same time. By presenting a completed design to the manufacturing staff, there is no opportunity to identify ways that, for example, simple design changes can improve manufacturing quality or decrease costs.

Using IPPD, manufacturers have the opportunity to find that new materials and processes can provide significant benefits. They can make these changes early in the design cycle when changes are made easily and inexpensively. As Figure 5–1 illustrates, these opportunities arise from viewing the design task in a larger context. This larger context includes geometry, materials, and processing as equally important design elements. The discipline of considering these three elements together can provide both cost and performance benefits.

A common application is the replacement of metal parts by redesigned but functionally equivalent plastic parts. The greater raw material cost for high performance plastic is offset by reductions in forming or assembly costs compared with the metal part. As an example, the bottom pan of a window air conditioner is traditionally

FIGURE 5–1
Design Approaches

Traditional Product Design

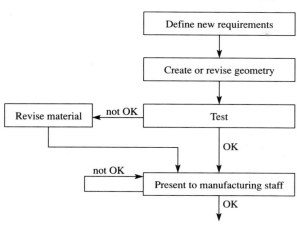

Integrated Product/Process Design

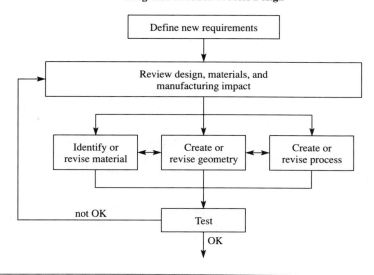

---

fabricated of sheet steel. This process involves inexpensive materials, but many fabrication steps. The pan is stamped, and brackets are attached to mount components such as the compressor. The mounting brackets must be formed, holes punched in the brackets and the pan, and the brackets attached to the pan. The entire assembly must be painted, with several coatings required to provide corrosion resistance and an attractive finish.

The comparable injection-molded plastic part requires more expensive raw materials. However, both corrosion and painting are eliminated, as are the stamping, forming, and assembly steps.

Implementing this material substitution requires major part redesign and process changes. Although the overall size of the pan would remain, as would the mounting locations for the components attached to the pan, the design of the plastic part would be quite different from the metal parts it replaced. The lesser physical properties of plastics compared with metal would require that molded-in stiffening ribs be added to the design to eliminate flexing and provide strength. The ribs must be arranged in specific ways to meet stiffness and strength requirements with minimum materials, tooling, and processing costs.

The manufacturing process would also be completely changed. Stamping, forming, riveting, and painting would be replaced by injection molding. Completely different tooling and production equipment would be used.

The payback for this investment in product and process redesign, as well as new tooling, is in per-piece savings and better product durability. But these benefits cannot be realized without simultaneous consideration of geometry, materials, and manufacturing.

The benefits of less-radical changes are also seen in examples such as galvanizing and advanced paint systems that provide better appearance and longer life for steel products. These changes may also require design, material, and processing changes. As an example, the use of rust-resistant galvanized steel car bodies affects the spot-welding processes used to assemble the body panels. The weld quality depends on the current flow through the electrodes, which in turn is controlled by the electrical resistance through the spot weld. Variations in galvanizing thickness cause variations in resistance and thus weld current. Uniformly good welds require close control of galvanizing thickness by the steel supplier and closer attention to welding electrode maintenance by the assembly plant.

Technology plays a critical part in the integration of geometry, materials, and processing. CAD provides the environment for the storage and retrieval of design data and thus helps enable the critical interaction among design, materials, and process engineering.

The integration of design, materials, and processing requires more than an appropriate environment. It requires expertise on the interaction among these three areas. This expertise is typically scarce. Manufacturers are challenged to acquire this expertise. Current staff, coming from specific disciplines, may not understand the interaction among areas. New materials and processes represent a moving target. Adding staff is often not an option, and so the alternative is to make better use of current staff by upgrading skills and providing new tools.

Knowledge-based systems have been shown to be useful in meeting these challenges, providing both the environment and the expertise to integrate and optimize across design, materials, and processing. This in turn has helped manufacturers realize the objectives of IPPD in ways not previously possible.

Upcoming sections describe applications of knowledge-based technologies to product design. Some of these applications call upon another advanced computer-based technology called **rapid prototyping.** This technology is gaining acceptance as another tool in the quest for more rapid design of higher-performance, lower-cost products. Although rapid prototyping is not a knowledge-based technology, it is a complementary technology. As described below, it is a useful manufacturing technology in its own right, as well as an element of a knowledge-based system solution for the design and fabrication of cast parts, as also described below.

5.2 RAPID PROTOTYPING

At some point in the design process, a physical prototype is required. A CAD rendering of a part is a good visualization aid, but there are needs addressed only by an actual 3-dimensional part. The prototype provides the first real sense of a part and how it will, for example, fit into the intended assembly. A prototype may also be required to begin fabrication of tooling.

Traditional prototype development methods are time-consuming. For this reason, the first prototype might be built well into the design cycle. Mistakes uncovered at this point are more expensive to correct than if done earlier in the design cycle, before other design decisions and constraints are imposed.

The goal of rapid prototyping is to save time in the design cycle and to identify problems when they are easily and inexpensively corrected. The approach is to develop a usable prototype part as rapidly as possible by generating a part directly from a CAD description of the part. Starting from a special type of CAD-based part description, a rapid prototyping machine will produce a part in a matter of hours, depending on the size and the required precision of that part.

The rapid prototyping output is a physical replica of the proposed part, but depending on the rapid prototyping process used, the prototype may be plastic, wax, or paper. The list of materials continues to grow in the interest of providing prototype parts that have usable physical properties as well as representative geometries. At a minimum, this part provides a reality test on whether design errors have been made and whether the actual CAD part description matches the intended geometry. This is a useful input before funds are committed to develop production tooling or to fabricate prototype parts in the material planned for production.

Regardless of the material, all rapid prototyping techniques build up the prototype parts as a succession of thin layers. The effect is that of assembling a loaf of bread by stacking individual slices one on top of another. For this reason the rapid prototyping process is driven by software that slices the CAD description of the desired part into a large number of pieces that, collectively, represent the desired geometry. The rapid prototyping machine successively generates these part cross sections and then fuses them into a whole.

The most common rapid prototyping process is called **stereolithography.** This process uses a liquid polymer that is transformed into a solid by ultraviolet (UV) laser light. The prototype part is produced by selectively solidifying the plastic to generate the desired geometry. Stereolithography equipment includes a vat of the initially liquid polymer and, within that vat, a vertically movable platform. Servo-driven optics enable the light from a UV laser to sweep the top surface of the polymer bath.

The stereolithography begins as the laser scans onto the surface of the polymer pattern that matches the shape of the bottommost cross section. This (partially) solidifies that cross section onto the movable platform, initially in the fully raised position. A wiper assembly passes across the top of the bath to smooth the plastic surface. The platform is lowered one cross-sectional slice thickness. The laser scans the shape of the next cross section, the wiper smooths the surface, and the platform drops another slice thickness. This process continues until all cross sections have been scanned, and a partially solidified but complete 3-D model has been created in the liquid plastic bath.

The stereolithography process subjects the laser to intensive use. Over time, the laser output diminishes, causing the rate at which the plastic solidifies to drop. Users balance the cost of replacing worn lasers against increasing part generation times. To conserve both the laser itself and the processing time, the liquid plastic is initially solidified only enough to allow the part to be self-supporting when removed from the liquid polymer bath. The polymerization process is completed off-line under ultraviolet light, producing a fully solid part.

To maintain part integrity and shape before solidification is complete may require that, for example, projecting part features be supported by ribs in the manner of flying buttresses on cathedrals. These ribs are added by the designer as part of preparing the CAD file for stereolithography. The ribs are then cut off the finished solid model. The designer might also add holes to the model to let liquid plastic drain from voids in the solid model.

Parts produced by this process are relatively fragile and can break if dropped onto a hard surface. Their ability to withstand dead, much less dynamic, loads is also quite limited. These drawbacks have led to interest in alternative rapid prototyping technologies.

As an example, DTM Corporation (Austin, Texas), offers the selective laser sintering (SLS) process. In this process, a scanning laser beam creates rapid prototype parts by sintering. Powdered materials such as polycarbonate, nylon, and wax are used and again fill a vat containing a vertically movable platform. The laser beam sweeps across the surface of powdered material and solidifies it by heating, and thus sintering, it into a solid. Again, the part is generated as a series of cross sections representing slices of the desired part.

The use of sintered wax in the SLS process offers the opportunity to create metal rapid-prototyping parts. This is done by a two-step process in which the SLS process creates wax molds that are then used in the investment casting process. This enables the rapid prototyping of parts that meet all functional requirements of production parts. In addition, this approach can be used to provide small lot sizes of production parts.

Investment casting is used to produce near net-shaped parts with complex geometries and tight dimensional tolerances. The process begins with the creation of a wax model of the part to be cast. Parts may be cast in groups, in which case wax models of individual parts are assembled in a fixture, and interconnected by wax runners. The wax is dipped into a ceramic slurry, which is then hardened to create a thin shell around the wax. The wax is melted out of the shell, leaving a void in the shape of the desired parts. Molten metal is poured into the shell, and the shell is broken away when the metal solidifies.

Manufacturers may pursue rapid prototyping in several ways. Some sophisticated and typically large manufacturers have in-house rapid prototyping equipment. Most call upon outside organizations that provide rapid prototyping services. These services may include, beyond creation of the prototype itself, preparation of the required computer files.

A CAD file is the required starting point and may need to be explicitly generated as a service for organizations that still do designs on paper. The CAD file is then transformed into a specialized form of geometry file, an **STL file.** The STL file contains a description of the surfaces of a part as a continuous surface made up of contiguous but separate triangular facets, each with specified vertices and a *surface normal,* a vector pointing in or out from the facet to specify whether that facet is on an outside or an inside surface of the part. Facet sizes are selected to provide an appropriate compromise between smoothness of curved surface representations and total number of facets and, thus, computational complexity and time. The faceted representation fully describes all points of the part's surface, as is required to unambiguously generate the part geometry.

An STL file is typically generated by a 3-D surface or solids modeling package, as may be found in sophisticated, typically workstation-based CAD systems. A surface modeler is used, for example, to verify that a part will fit into an assembly without interference with adjacent parts. A solids modeler might be used to determine that a part will provide acceptable dynamic performance.

Many manufacturers employ such modeling tools and therefore already have the capability in house to generate STL files directly. Other manufacturers using less-sophisticated design tools may choose to implement 3-D solid modeling software specifically to provide the STL files required for rapid prototyping. This will require designers to learn to use a new design package and may also require the purchase of a workstation and peripheral equipment. Or, manufacturers may choose to avoid these up-front expenses and training costs by purchasing the generation of STL files as a service.

The final software step in preparation for rapid prototyping is to use a slicing routine to transform the design of the part as a whole into a collection of slices. The slice thickness is user-selected. Thinner slices require more computation time and, more important, more time to generate the rapid prototyping model. Thicker slices provide a more pronounced sawtooth effect on sloped surfaces. This sawtooth effect on sloped lines is due to step-wise approximations of smooth lines, with the size of the step representing a trade-off between speed and accuracy.

An application of rapid prototyping is provided by Linear Solutions, a start-up company designing and building auxiliary lighting systems for ambulances and police cars. This market is dominated by a large, 70-year-old competitor, and so Linear Solutions had to rapidly and effectively design a new product.

That product was a spotlight for emergency vehicles. The spotlight had to be attractive, while blending in with the smooth shapes and compound curves now used to make vehicles more aerodynamic. The light also had to be sturdy enough to withstand rough service. These requirements imposed design challenges. The individual parts were of complex shapes, difficult to represent on a standard 2-D drawing. The individual parts had to fit mating parts in the assembly to produce an assembly that was attractive and optically effective.

These challenges required Linear Solutions to develop prototypes before committing to the final tooling. The prototypes were to be of metal to enable rigorous testing. The traditional approach would be to machine aluminum prototype parts. This would be an expensive and time-consuming process made more so by the machinists' difficulty in interpreting drawings of the complex shapes.

Linear Solutions pursued an alternative, rapid prototyping. They chose the SLS process because it enabled parts to be investment cast in aluminum. The DTM Sales and Service Center began by translating the 3-D wireframe model designs provided by Linear Solution into the required format. DTM delivered the six prototype parts, the largest of which measured 8 inches by 5 inches by 3 inches, in six weeks.

Linear Solutions estimated that the SLS-generated parts were delivered in half the time required by traditional methods and at one-half to one-third the cost. The availability of prototype parts was also reported to enable the designers of the final part tooling to identify simpler approaches, trimming $20,000 to $30,000 from the anticipated $150,000 cost of tooling.

A second rapid prototyping application will be described as a part of the upcoming section on the Rapid Foundry Tooling System. In this application, stereolithography and knowledge-based technologies are used to rapidly design both parts and tooling for aluminum sand castings. Knowledge-based technologies are used to identify design and tooling strategies and potential problems and solutions for each. Rapid prototyping is used to validate the designs.

The following section returns to the primary focus of this chapter: applications of knowledge-based technologies for the design of discrete parts.

5.3 EXPERT SYSTEMS

Many design tasks involve an element of creativity, but most require that rules be followed. These rules may ensure that functional needs are met, that production feasibility is addressed, or that current practices are observed. One way that an expert designer differs from a novice is in understanding the existence of these perhaps implicit rules and their impact on product design.

Rule-based systems have been widely applied to provide this type of support in the design of discrete parts, as the following applications illustrate.

5.3.1 Shipbuilding

Ship design is a complex task involving complex geometries as well as shape, size, weight, and strength requirements. These capabilities cannot be compromised by the holes required to run plumbing, wiring, and ductwork through the ship structure. Ingalls Shipbuilding is reported to have developed a Penetration Control Advisor to help design engineers meet potentially conflicting requirements. The system provides advice regarding structural integrity and reinforcement requirements to preserve structural strength when providing the holes in the ship structure required to accommodate wiring, plumbing, and ventilation ductwork.

5.3.2 Nuclear Reactor Design

Lummus Crest, a subsidiary of Combustion Engineering is reported to use a design system called Design of Reactors in Space (DORIS). This system is integrated with a CAD system and assists a complex packaging task. That task is to find a location and shape for the several miles of metal heat transfer tubing that must fit into a confined space without interference with other components. DORIS is claimed to reduce reactor design time from eight months to eight weeks.

5.3.3 Software Packaging

Anderson Consulting developed a system for Microsoft Corporation called Bill of Material Workbench. This system is reported to help engineers meet guidelines in the number of software diskettes, the packaging of these disks, and pricing of the product. The system accesses CAD information as well as cost data and current bills of material residing on remote systems. Benefits claimed include faster and more accurate design and more consistent application of guidelines from a central source.

5.3.4 Helicopter Design

McDonnell Douglas Helicopter Company is reported to use an expert system to support the design of composite parts. The system helps calculate the part size requirements and from this determines the minimum weight and the number of plies of composite material required. Identified benefits include reduction in design time and the ability to perform part weight-versus-strength analysis earlier in the design cycle.

5.3.5 Boiler Design

Babcock & Wilcox has used the Concept Modeler shell, developed by Wisdom Systems, to perform similar design and configuration tasks for fossil-fuel–fired boilers as used in large-scale electric utility power generation. The design of these boilers requires that heat exchanger tubing, combustion air fans, burners, and a variety of other design elements, all of which involve complex geometries, be combined into an integrated design. Boiler design and configuration rules assist engineers in meeting functional and packaging requirements while observing geometric constraints and material selection requirements for individual components.

Stone & Webster Engineering has developed another boiler design system called STONErule to support their industrial applications. This system incorporates electrical and electrical design knowledge, as is used in the design of power generation equipment. STONErule functions as an integral part of a CAD system. Engineers working in the CAD system pose questions to and receive answers from the expert system without leaving the CAD system or being aware that the expert system exists as a separate entity. Users may incorporate their own knowledge in plain-English rules rather than an AI language such as LISP. This makes system maintenance easier and improves the likelihood that the system is provided with up-to-date knowledge.

5.3.6 Environmentally Controlled Structure Design

Bally Engineered Structures, Inc., designs and builds specialized structures, including insulated outdoor structures, refrigerated warehouses, walk-in coolers and freezers, and environmentally controlled rooms. These units can be configured in virtually limitless combinations of standard elements to meet specific needs.

Bally began in 1933 as a job-shop manufacturer. Each order was custom-engineered and production was handled on an order-by-order basis. As the industry matured and competition intensified, Bally evolved from custom engineering to mass production. Although individual jobs were still customized, they were assembled of standard modules. In some cases, individual modules were modified as required. In other cases, individual jobs were distinguished only by configuration of the standard modules.

Bally was purchased by Allegheny International, and in 1983 a new CEO was brought in. He found that declining revenues were caused not only by market saturation and resulting price pressures, but also by stagnant product designs, obsolete processes, and excessive administrative costs. He determined that Bally must make radical changes and that these changes must focus on administration more than manufacturing as such.

To provide more competitive products through greater customization, the Bally line grew to 10 times the previous number of combinations. The complexity and possibility for error in the design process increased a like amount. Each order was also revised an average of two and a half times before fabrication was completed. The result was that a structure requiring four days to build required two and a half weeks for initial design, and up to seven weeks to process change orders.

The change orders represented significant administrative (not to mention scheduling) costs. For this reason, an integrated computer-based administrative and manufacturing information system was developed. It incorporated an expert system for product configuration.

The expert system for product configuration contains the knowledge of Bally configuration experts. This system was shown to provide consistently accurate results, and eliminated the need to check and recheck the product configuration. Reducing configuration checking and the processing of change orders that resulted from the discovery of errors provided significant time savings. Now, the overall design/configuration process is reported to typically require less than four hours.

5.3.7 Automotive Air Conditioner Design

General Motors Corporation has reported the development of a tool to assist the design of automotive air conditioning systems. This tool, called the HVAC (Heating, Ventilation, Air Conditioning) Design Advisor, uses the Knowledge Engineering Environment (KEE) tool from Intellicorp. The HVAC Design Advisor incorporates both design calculations and problem-solving advice.

Customers expect automotive air conditioning (A/C) systems to rapidly cool the vehicle and maintain comfort on the hottest day. Maintaining this level of performance is becoming more challenging as less efficient but environmentally benign Freon substitutes are implemented and as vehicle designs call for increased glass area and, as a result, increased solar heat loading.

The A/C system for a new car is best designed early in the overall design process because two of the system components, the condenser and the evaporator, are relatively large. They must compete for space in the densely packed engine compartment and must be provided sufficient clearance for ductwork to carry the required airflows. The compressor must also find room on an increasingly crowded engine.

These packaging issues require that setting design parameters for a new car be preceded by an estimate of air conditioner component requirements. The traditional approach was to build a working prototype with a new compressor, evaporator, and condenser installed in a current car closest in specification to the anticipated new design. The performance was measured to determine whether overall comfort objectives were met.

This approach was expensive and time-consuming. As new car designs evolved, the cooling requirements and packaging constraints might also change and make tests performed on the A/C system prototype no longer representative of the anticipated performance.

These factors provided the rationale for development of the Design Advisor. The Design Advisor calculates cooling loads and designs equipment to meet these requirements. The Design Advisor also incorporates expert knowledge from within General Motors to help check whether design parameters are outside normal ranges and to suggest solutions to performance and configuration problems.

A design is developed with the help of the Design Advisor and its two elements, Design Aide and Design Critic, as follows:

- The user selects a current car closest to the anticipated new car. The Design Aide then provides design information on that car and its A/C system.
- The system calculates the solar heating load for the new car based on glass surface areas, orientation, and curvature of each window, as well as the optical properties of the glass and the resulting heat transfer by conduction, convection, and radiation.
- The A/C system capacity is calculated to meet quantitative system performance requirements and qualitative measures of perceived comfort.
- An iterative approach is used to design the compressor, evaporator, and condenser to best meet performance requirements within constraints of components' size, weight, durability, and cost. A traditional thermodynamic analysis package is incorporated into the Design Advisor's expert system framework.
- The Design Advisor also includes a so-called Design Critic. Knowledge of current practices and normal ranges of design parameter values is used in the background to validate decisions as they are made. Deviations are brought to the user's attention. For example, a warning might be given that the output pressure of the compressor exceeds the normal upper limit established on the basis of durability and performance requirements.

In the Design Aide role, the system suggests specific design changes to resolve specific problems such as this. For example, the system may suggest several methods to provide better condenser performance, which in turn would be one of several ways to increase overall system performance.

The Design Advisor is reported to be used to make early design-stage decisions and reduces the need for A/C prototypes. It provides a platform to store in-house expertise and a rationale to collect and organize that expertise before it is lost.

The Design Advisor also illustrates a trend in knowledge-based systems to combine a variety of technologies, knowledge-based and traditional, into a single integrated system. In this case the

Design Advisor incorporates thermodynamic, graphical analysis, and other analytical techniques in addition to the knowledge-based elements. This provides a more appealing problem-solving environment because the system supports a greater portion of the design task and minimizes the need to repeatedly exit one program and enter another. The matter of user appeal cannot be overemphasized. Computer-based systems provide no benefits if they are not used, and they will not be used, at least to their full potential, if users do not find the system to be both helpful and comfortable.

5.3.8 Computer Configuration

One of the most celebrated expert systems is XCON by Digital Equipment Corporation. As one of the first commercially significant expert systems, XCON is useful as a benchmark. XCON is an eXpert CONfigurator of computer systems and was built to support the configuration of VAX supermini computers. The task was to select, in response to a customer order, central processing units, processor options, memory modules, tape drives, disk drives, adapters, cables, terminators, bus repeaters, power supplies, cabinets, terminals, and software. From 10 to 200 parts would be selected for an order from the 8,476 available items. Millions of combinations were possible, and systems were generally customized to some degree. The customer order would be checked for completeness and functionality. Compatibility with other hardware had to be enforced, and all components had to physically fit the available cabinet space. (In 1985, a supermini computer such as a VAX 11/782 came in a cabinet perhaps 15 feet long and 6 feet tall.)

XCON is still in use. Sales people use it to help prepare quotations. Manufacturing schedulers use XCON to verify that only feasible configurations are scheduled for production. Assemblers on the Digital factory floor and technicians at customers sites use complex configuration diagrams generated by XCON as an assembly guide.

Work began on XCON in late 1978. The expert system approach was selected to facilitate detail changes, as well as major restructuring of XCON to reflect major product and configuration changes. A 250-rule prototype was demonstrated in April 1979. Testing on 50 sample configurations lasted through November 1979. Correction of errors and gaps in the rule base expanded XCON to 750 rules. XCON was installed at Digital manufacturing sites throughout 1980. From then until 1983, XCON was rewritten twice into more current programming languages and extended to support more products. A shell was added to surround the XCON database and to provide performance monitoring and problem reporting capabilities. By early 1985, XCON had about 4,200 rules.

Digital paid attention to system usage and maintenance issues. A 14-week training program was initiated in January 1983 to develop in-house skills in the development and maintenance of expert systems including XCON. This enabled Digital to gradually take over all aspects of XCON maintenance without further reliance on members of the original development team from Carnegie-Mellon University.

Digital's long involvement with expert systems provided an environment as receptive as any to the introduction of this new technology. Even so, XCON developers had to overcome skepticism among potential users. They did this by demonstrating the accuracy and benefits of the system. XCON's initial accuracy was reported to be 75 percent rising to 95 percent over the next 18 months while accommodating new product and configuration rules. Digital reported that XCON accomplished the average configuration in approximately two and a half minutes, performing about six times as many functions as were done manually; as of 1985, XCON achieved feasible and complete configurations 98 percent of the time. Before XCON, people performed system configurations in an average of 30 minutes and got 70 percent of them correct. (These times do not apply to the more complex VAXcluster configurations.)

More recent reports on XCON show continuation of prior trends. In 1987, XCON was reported to have processed 80,000 orders. In 1989, a far more complex XCON consisting of 10,000 rules was able to configure more complex systems, now consisting of 200 to 8,000 parts. In 1991, XCON was reported to save $200 million per year but to cost $2.5 million to maintain.

These numbers cause some observers to debate the merits of XCON in particular and expert systems in general. Is the claimed $200 million savings in comparison with a nonexpert system configuration tool, or is it in comparison with hiring the many additional people required to configure manually? Considering the $2.5 million maintenance cost, one might ask if the anticipated benefits of easier maintenance and modification have been demonstrated. Of course this question can never be directly answered because no one is going to redo XCON or any other complex system another way to provide the direct comparison.

One might also speculate that the big numbers associated with XCON have more to do with the configuration task itself than with the expert systems technology. A lesson common to many automation tasks may apply here. Perhaps 80 percent of the benefit of a successful automation project is derived from better understanding a process and then rationalizing and simplifying that process. Only 20 percent of the benefit is directly attributable to the automation itself. In the case of VAX configuration, it may be that the benefits of XCON are overshadowed by a massive configuration task.

This is speculation and may have no merit in the case of XCON. But there appear to be a number of conclusions that can be drawn from Digital's pioneering expert system applications:

- Expert systems technology has demonstrated problem-solving capabilities in support of a variety of design tasks, particularly those in which clear-cut choices can be made according to explicit and relatively unchanging rules.
- System development can be a major undertaking. System development periods of two years are common and often require multiple experts and one or more knowledge engineers and programmers.
- Maintenance is a significant task that must be explicitly addressed. It requires considerable time and skill, even under the best possible circumstance of continued access to the original system designers and domain experts.

Since XCON, expert system development and maintenance have been made easier with commercial development shells that provide a predeveloped user interface and rule-processing engine and that are programmed in English. Nonetheless, the development and maintenance issues surrounding expert systems have driven the definition and development of alternative knowledge-based technologies.

The fundamental need is for easier ways to acquire knowledge. The appeal of one of these alternatives, neural nets, is the ability to discover relationships within data and, in this way, discover knowledge. The following example clarifies some of the benefits and issues associated with this approach.

5.4 A NEURAL NET APPLICATION: ALUMINUM SANDCASTINGS

The foundry at the Air Logistics Center of Kelly Air Force Base (San Antonio, Texas) provides aluminum sandcastings as replacement parts for a wide variety of equipment. Kelly is responsible for all aspects of production, including design work as may be required, fabrication of molds and other specialized tooling, and production of the castings themselves. Small-order quantities are typical, but rapid delivery is important.

Kelly shares many of the challenges common to most casting foundries. These include the need to provide high-quality parts through careful attention to part and process design and to accommodate the wide range of designs and complex geometric features that characterize castings. The sandcasting process is as much an art as a science. The decisions made in the design of the tooling for a particular part reflect personal experience and preferences as much as hard-and-fast rules.

Kelly also encounters unusual challenges. Kelly must supply replacement parts for specialized equipment that may have been built many years ago, in small quantities, and for which spare parts are no longer available. Beyond this, part drawings may no longer be available. In this case, the first step must be to **reverse engineer—** that is, recreate the design of the castings.

The need to streamline the fabrication of aluminum sandcastings caused the US Air Force to work with AIWARE, Incorporated, in the development of a unique knowledge-based design support system. The Rapid Foundry Tooling System (RFTS) incorporates neural networks and an advanced computer memory technology called *episodal associative memory* to accomplish the following:

- Reverse engineer a casting, producing a CAD description.
- Validate the design with rapid prototyping.
- Ensure that both part and process designs reflect good

practice to minimize likelihood of defects and support rapid delivery of parts.

- Capture the expertise used to accomplish these tasks, including the lessons learned, so that future design and fabrication tasks can draw upon prior experience.

The functions of RFTS and the technologies enabling these functions are described below.

5.4.1 Reverse Engineering

The usual design task begins with definition of design and identification of geometric and functional constraints. Designs are developed, and parts are built and integrated into an assembly or some larger environment.

However, circumstances arise in which part design cannot be done this way. The maintenance of equipment can be such a circumstance. The environment, the assembly, or the machine already exists, but for some reason, the design of an individual part is no longer available. The part supplier may no longer be in business. Drawings of CAD files may simply be lost. In these cases, reverse engineering may be employed.

Reverse engineering may not be appropriate in all cases. One must be sure that the equipment to be maintained is worth the expense of reverse engineering and fabrication of parts. However, in both defense and commerical applications, cost may not be the dominant issue. An immediate need for a piece of equipment may justify reverse engineering, as might the costs and bother of replacement or the absence of alternative equipment for specialized tasks.

Traditional methods of reverse engineering are those of standard engineering but applied in reverse order. To best meet rapid delivery requirements, the Air Force has pursued nontraditional methods of reverse engineering. A laser-scanning coordinate measuring machine makes best use of available information when a worn or broken part is available. Part geometries are generated by the scanner and transferred to the CAD system to provide at least a portion of the part geometry. The designer completes the design rather than starting from scratch.

The more common coordinate measuring machines use a hard-tipped stylus that is moved through space to touch various points on the surface of the part being measured. The result is a table of 3-D coordinates representing points on the surface of the part.

The alternative approach employed by Kelly Air Force Base is to use reflected laser light to generate a continuous 3-D surface representation of the part. A laser light beam is swept across the part surface. A camera detects the reflected light. The location at which the reflected light is detected depends on the location from which it is reflected. The laser scanner thus uses triangulation to generate a database of the complete 3-D surface. The laser-scanned approach is faster than the hard-stylus approach, although with both approaches, the length of the time required is based on the measurement resolution requirements, which in turn determine the spacing between adjacent probes or scans.

RFTS supports reverse engineering by

- Translating the output of the laser scanner coordinate measuring machine into a CAD format, eliminating manual translation from one format to another.
- Translating the output of the CAD system into an STL file, to drive stereolithography rapid prototyping equipment.

In both cases, these translations add some information, but mostly they overcome incompatible file formats. A reality in the integration of computer-based systems is that rarely can one computer system talk directly to another without the use of a translator called a *bridge*. There is rarely a fundamental reason why this is so, other than that different application areas, such as CAD and CNC, and different companies employ equally valid but arbitrary and different choices for computer file and communication formats.

The trend is in the direction of more universal file formats to minimize the need for software format translators. These efforts have proceeded slowly, and one can envision disincentives to develop universal formats (read: the retention of customers and higher margins that come with proprietary systems). In the meantime, software developers accomplish translation on a case-by-case basis, a workable if not particularly elegant approach.

Returning to casting issues, the challenge in computer-based support of castings is that there are many right ways to design a cast part and its associated process. For this reason, a casting support system needs to provide problem-solving advice but to not arbitrarily preempt any approach. If the system suggests a single solution to a given problem, those who have successfully used other approaches will dismiss the computer-based system as incomplete or irrelevant.

RFTS merges part design, materials optimization, and processing design with a modeling capability that spans these three domains. RFTS draws on prior experience and relationships discovered between casting design and processing. This enables users to review and choose among different design and manufacturing strategies by identifying the potential problems and solutions associated with each.

The knowledge-based technologies in RFTS are based on a neural net-based modeling technology, integrated with an optimizer, as incorporated in AI WARE commerical products. In RFTS, this general-purpose capability was extended to provide a unique design support capability with two components:

- Design and manufacturing experience is captured as *episodes* rather than as rules or data.
- This information is used to find the prior design and manufacturing episode closest to a desired new casting and to learn by example to identify good design practices and potential problems.

These capabilities are enabled by a technology called **episodal associative memory** that is based on neural networks but conceptually may be viewed as an extension of expert systems.

5.4.2 Episodal Associative Memory

A major practical problem in knowledge-based systems is that of collecting and updating knowledge and organizing that knowledge so that it can be easily and effectively reused.

Expert systems and neural networks represent two approaches to the problem. each with strengths and limitations. Expert systems provide specific responses to predefined problems. Their ability to support product design is limited by their inability to generalize, to discover how prior designs apply to a new and different design. Neural nets generalize, but it is a challenge to use numerical inputs and outputs alone to describe the discrete choices to be made in the design of discrete parts.

There is evidence that one way people remember large bodies of varied information is to treat this information as a collection of episodes. Within each episode, much information is captured and remembered on the basis of its association to the overall theme of the episode.

As an example, the theme "my first job" conjures up a wealth of information, much of which has to do with the job and much of which relates to peripheral details: the office setting, co-workers, and even the season. Further episodes, such as other jobs, may be remembered as deviations from the first job. Information is stored more compactly and accessed more readily when captured as a nominal situation and deviations from nominal rather than as a number of complete but unrelated bodies of information.

In the same way, the designs of cast parts are similar in many ways, especially when those designs are divided into classes. Beneath these common themes lie detail differences. Conclusions can be drawn from both the similarities and the differences.

Dr. Yoh-Han Pao developed the concept of the episodal associative memory (EAM) as a computer-based technology that is based on that model of learning. The EAM captures information in the form of descriptions of specific episodes. The information is retrieved and made available to solve new problems on the basis of an association drawn between the current problem and relevant prior problems.

This association to a specific episode, and the wealth of information associated with that episode, enables the EAM to provide the user the focused, clear-cut response typically associated with an expert system. The EAM can call upon its memory of data, advice, rules, and lessons learned. It can also explain the basis of the answer by identifying the episode to which the current situation is believed to be most closely associated.

The neural net underlying the EAM learns relationships automatically, enabling the system to generalize and improve its own performance. In RFTS, this generalization capability is used to discover which prior design a proposed design is closest to. For a system such as RFTS that is intended to support a wide range of designs, explicitly programming each design or family of designs could be an unmanageable task. Instead, RFTS incorporates a capability that can be viewed as an automated implementation of group technology.

In group technology, parts are divided into families or groups on the basis of similarities in design features and processing steps. Outside diameters within a certain size range and within a range of surface finishes might be identified as members of a group so that these parts could be machined together and thus minimize the frequency of machine setup. The underlying premise is that the time spent by people analyzing designs to identify similarities is more than recovered on the factory floor.

EAM also identifies similarities but automatically, as follows:

- EAM uses a neural net-based model to find multidimensional similarities. Parts are judged to fit into classes not on the basis of individual features and dimensions but on the basis of combinations of features, dimensions, tolerances, processing, and materials.
- The neural net-based model makes this determination automatically, calling upon the clustering capability of unsupervised learning. This saves time and minimizes the need for design and processing expertise for the design and maintenance of RFTS and other systems.

The EAM-based RFTS provides specific design recommendations similar to those that might be provided by an expert system, even though the EAM is an extension of the distinctly different neural net technology. The recommendations provided by RFTS include identification of the most relevant prior specific designs in ranked order of similarity. Processing problems associated with each design and associated solutions are also identified.

5.4.3 RFTS Knowledge

In concept, casting can be viewed as making ice cubes of metal: pour liquid metal into a form, let it solidify, remove the form. In practice, the situation is more complex because a cast part typically has features on all surfaces, unlike flat-topped ice cubes. This means that the part mold must be at least two parts that together provide a completely enclosed void in the shape of the part, with the only openings being one or more small gates through which liquid metal is introduced.

Once the part is cast, it must be removable from the mold without damage. This requires that the part incorporate tapered surfaces so that, as with ice cubes, the part slides easily out of the mold. The tapered surfaces meet at a *parting line,* which represents the interface between the two halves of the mold and the finished part. The slight leakage of liquid metal through this interface creates a ridge on the part surface.

Placement of the parting line on the part may be an important decision and may involve complex geometries. The parting line should be placed so that the ridge is not objectionable functionally or cosmetically. Placement of the parting line must also allow the two halves of the mold to be assembled together without interference or damage.

The geometry of some parts precludes a two-piece mold. In these cases, additional pieces known as *cores* are required to define, in the mold, the shape of the part without preventing disengagement of the mold from the completed part. It can be a complex problem in 3-D geometry to design a mold (and perhaps cores) in a way that provides the required part features and allows assembly of the mold before casting and removal of the finished part after casting.

Other mold design issues may not be directly visible in the finished part. The parts may exhibit porosity or incomplete features if, for example, liquid metal is not provided to all areas of the mold as solidification occurs. This requires careful design of the mold features that bring liquid metal to all points within the mold.

Molds incorporate a system of channels called *rigging* through which the molten metal runs from the pour point to the location of the finished part. The design of the rigging and the orientation of the part itself in the mold can affect product qualtiy. The goal is to ensure that liquid metal is able to completely fill the mold before it solidifies. If not, the part will be incompletely formed.

Runners, the passageways composing the rigging, must be of sufficient size. Small reservoirs called *risers* may also be required near certain part features to ensure that liquid metal is available to fill voids presented by shrinkage during solidification. Finally, the size and location of gates—the points at which metal leaves the rigging and enters the mold—must be designed to ensure that metal does not freeze at a thin cross section of the part and fail to fill a mold section beyond that point.

In principle, this information could be provided by an expert system. In practice, this would be difficult. Since liquid metal must flow by gravity to completely fill the mold, it is evident that the size, orientation, and relationship of each element of the part and the rigging will affect mold filling and thus part quality. It would be difficult to describe in rules the acceptable relationships between various design features for the wide variety of parts produced by Kelly. Since many of these parts cannot be anticipated during system development, it is not possible to ensure in advance that an expert system will cover that future part.

In addition, each expert moldmaker has a preferred approach, and in most cases several approaches will work equally well. One challenge in supporting rigging design is to identify potential problems and solutions without arbitrarily precluding a moldmaker's favorite approach. EAM enables RFTS to find those designs, problems, and solutions most relevant to a current design task, so that the user can avoid problems but not be unnecessarily constrained in approach.

5.4.4. RFTS in Use

The user will start with a CAD description of the proposed part. This description may be generated through a traditional design process or through reverse engineering. Figure 5–2 illustrates the RFTS problem-solving modules for a typical part—a bearing housing. The initial CAD description is shown in the upper left.

The user then calls upon the **feature-based design** environment of RFTS as shown in Figure 5–3. The user identifies part features such as bosses and through holes and, on this basis, EAM identifies those parts that provide the closest match based on geometric features. These candidate parts are listed in the lower right corner of Figure 5–4. The user may review each of these to identify favored approaches and verify closest match to the current design requirement. The user has selected one, cannister 3, for display. This part is displayed on the right side of Figure 5–4 along with the associated rigging to provide a suggested rigging approach for the bearing housing. Although the cannister on the right and the bearing housing on the left side of Figure 5–4 are dissimilar in detail, EAM has determined that similarities between the two justify consideration of the rigging arrangement shown.

To identify potential rigging-related problems and solutions, the user may refer to Figure 5–5, which illustrates the closest EAM-

FIGURE 5-2
RFTS Modules

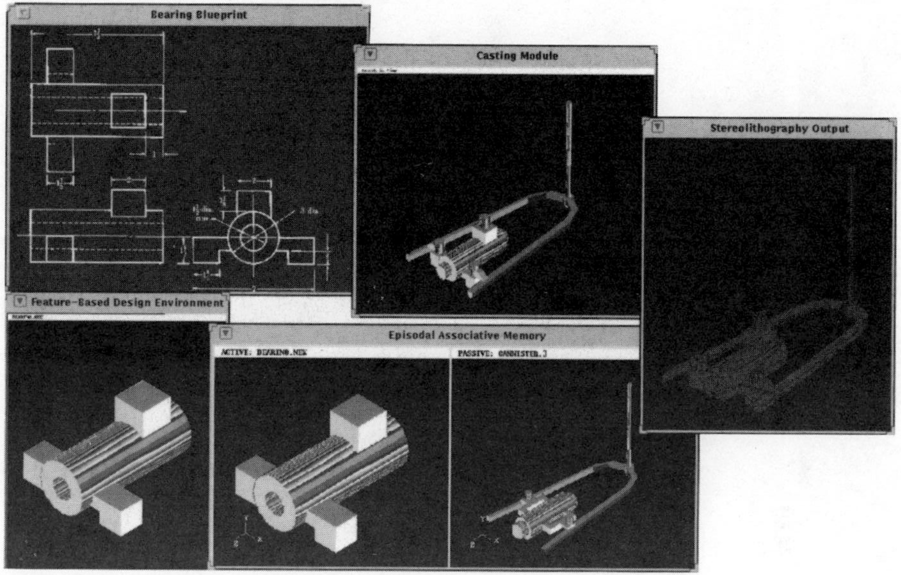

FIGURE 5–3
RFTS Feature-Based Design

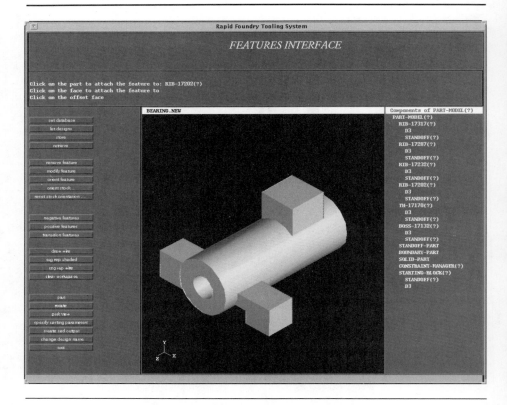

identified match of rigging arrangement. The design of bearing 9 shown on the left side of Figure 5–5 is not particularly similar to bracket 4 on the right. But the rigging arrangement is the one identified by the EAM to be the most similar rigging arrangement within its memory, based on the neural net-enabled match of rigging (not part) features.

The user then learns from Figure 5–6 that this part—bracket—suffered from shrinkage when cast with this particular rigging arrangement. The user will want to know more about the problem and its solution, and so will call upon Figure 5–7 to find that the EAM identified the underlying problem to be a part section

FIGURE 5-4
EAM for Most Similar Parts

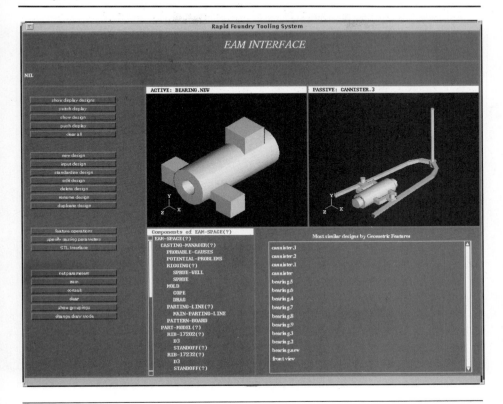

too heavy to be fed by the available rigging. The user then learns through Figure 5-8 that the addition of *chills* solved this problem. Chills are heat sinks that, when inserted into the mold in the proper location, provide a temperature gradient that changes the pattern of metal solidification and so prevent voids in the finished part.

The user may accept the recommendations presented or try a completely different approach, calling upon alternative part or rigging design strategies based on the next closest alternatives identified by the EAM.

Once a design has been established, the user may use the stereolithography module. This module transforms the CAD description

FIGURE 5–5
EAM for Most Similar Rigging

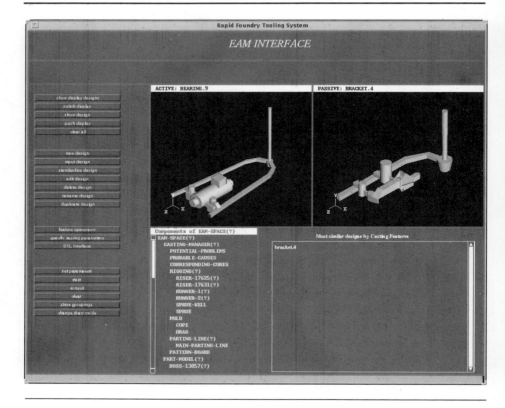

into an STL file, as shown on the right side of Figure 5–2. This module also contains knowledge to best accommodate stereolithography capabilities by

- Minimizing part height (and therefore processing time).
- Optimizing part draining.
- Minimizing the number of slanting and sloping surfaces (to minimize the resulting rough surfaces).
- Maximizing the number of smooth surfaces facing vertically and horizontally.
- Ensuring that the part will fit into the resin vat.

FIGURE 5–6
EAM for Problem Identification

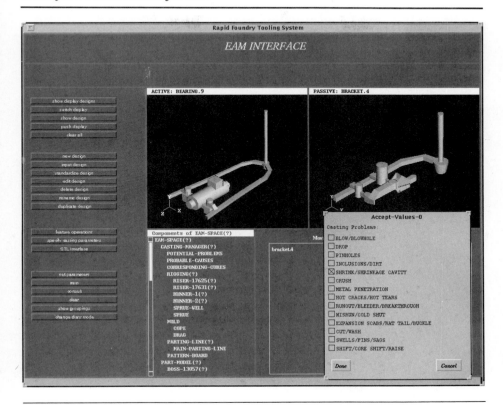

Since the designer must often provide supports for the partially solidified model produced by sterolithography, RFTS helps the user by

- Deciding where to use supports in the form of webs, solid cylinders, hollow tubes, triangular columns, and square columns.
- Ensuring that the supports overlap the objects but do not intersect cosmetically or dimensionally important areas of the part.
- Providing supports that are easy to separate from the object

FIGURE 5–7
EAM for Problem Characterization

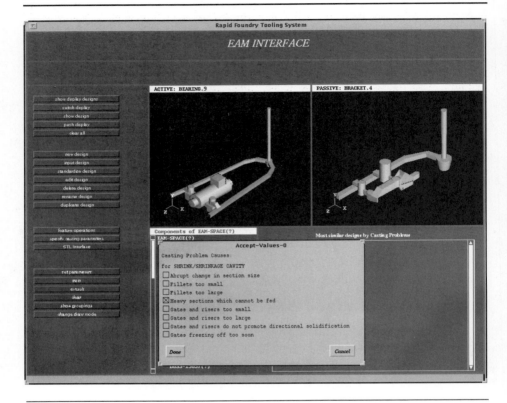

and fewest in number but accomplish the goal of preventing the part from bending.

- Providing cross supports to prevent the part from tilting and skewing.

RFTS also helps the user manage the faceted part representation by

- Providing facet normal information, indicating the direction of surface curvature (enabling the user to verify that all are correctly oriented).
- Helping the user generate a closed surface.

FIGURE 5–8
EAM for Problem Solution

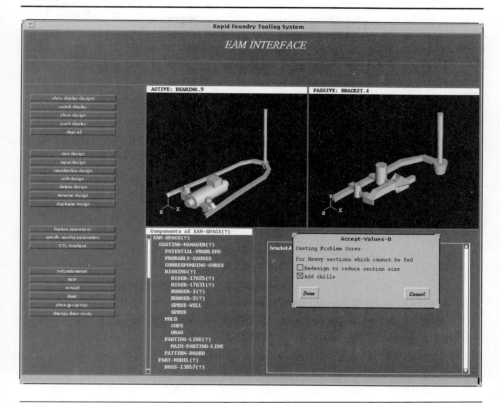

• Adjusting the facet resolution (size) to provide the required accuracy of representation and smoothness of surface.
• Providing sufficient wall thickness.

The information within RFTS was acquired from casting designers, moldmakers, and casting handbooks. Although RFTS was developed to meet the needs of the foundry at Kelly Air Force Base, EAM represents a general-purpose technology with wide application potential. Variations on the RFTS theme could be developed for any manufacturing process that is sufficiently complex or dependent upon special expertise to justify the development of a sophisticated computer-based system. Such tools would be useful as training tools as well as real-time decision-support tools. Application areas might include related manufacturing processes such as injection molding.

The preceding applications illustrate many of the ways that knowledge-based technologies support the design of discrete parts. Expert systems capture design expertise to enforce functional and manufacturing requirements. Neural nets discover design relationships, and the episodal associative memory identifies potential design and manufacturing problems and solutions. These capabilities extend computer-aided design in ways that help manufacturers provide better products faster.

Similar benefits have been demonstrated for products that are produced by batch and continuous processes. The next chapter includes an overview of design issues for these products. Knowledge-based applications demonstrate the benefits of these technologies for this different class of products.

5.5 REFERENCES

1. Inglesby, Tom. "Industry Profile: An Interview with Larry Ford." *Manufacturing Systems,* October 1992, pp. 34–38.
2. Kestelyn, Justin. "Application Watch: Smart Product Configuration." *AI Expert,* July 1990, p. 72.
3. Kestelyn, Justin. "Application Watch: STONErule Measures Up." *AI Expert,* April 1992, p. 64.
4. Keyes, Jessica. "Manufacturing Survey." *Handbook of Expert Systems in Manufacturing* New York: McGraw-Hill, Inc., 1991, pp. 42–55.
5. Keyes, Jessica. "Why Expert Systems Fail." *AI Expert,* November 1989, pp. 50–53.
6. Manji, James F., ed. "Slashing Time and Money from Art to Part." *Controls & Systems,* June 1992, pp. 32, 34.
7. "NEXPERT OBJECT in Manufacturing and Design." Distributed by Neuron Data, Palo Alto, CA: October 1990.
8. Pao, Yoh-Han. "The Epidosal Associative Memory: An Approach to the Management of Manufacturing Information." Technical Report

TR 90–136. Cleveland, OH: Center for Automation and Intelligent Systems Research, Case Western Reserve University.

9. Pao, Yoh-Han, K. Komeyli, D. Shei, S. LeClair, and A. Winn. "The Episodal Associative Memory: Managing Manufacturing Information on the Basis of Similarity and Associativity." *Journal of Intelligent Manufacturing,* 4(1993), pp. 23–32.

10. Pao, Yoh-Han, F. L. Merat, and G. M. Radack. "Memory Driven Feature-Based Design." Report WL–TR–93–4201. Dayton, OH: Materials Directorate, Wright Laboratory, Air Force Materiel Command, Wright-Patterson Air Force Base.

11. Pine, B. Joseph, II, *Mass Customization: The New Frontier in Business Competition.* Boston: Harvard Business School Press, 1993, pp. 143–45.

12. Ramus, Dan. "AI in the '90s: Its Impact on Manufacturing." *Manufacturing Systems,* January 1991, pp. 32–39.

13. Scown, Susan J. *The Artificial Intelligence Experience: An Introduction.* Maynard, MA: Digital Equipment Corporation, 1985, pp. 111–25.

14. Stanke, E. J., A. Bajpai, B. M. Boone, G. E. Clark, J. F. Gratton, and R. J. Niemiec. "A AI Tool for Flexible Design of Automotive Air-Conditioning Systems." Detroit: IPC '92 Proceedings of the ESD International Programmable Controls Conference, ESD—The Engineering Society, 1992, pp. 643–51.

15. VerDuin, William H., and Yoh-Han Pao. "The Rapid Foundry Tooling System: A Cutting-Edge Computer-Aided Design System." Dayton, OH: Proceedings of the IEEE 1993 National Aerospace and Electronics Conference, vol. 2, pp. 926–29.

Chapter Six

Knowledge-Based Formulated Product Design:
Better Properties and Lower Costs

6.1 INTRODUCTION

Chapter 5 presented discrete part design problems and knowledge-based solutions. Knowledge-based systems were shown to help manufacturers meet new marketplace requirements rapidly and effectively.

The alternative to discrete parts are products made by continuous or batch processes. Materials are combined and transformed by processes that may deal with individual batches ranging in volume from milliliters to railroad carloads. Or these materials may be transformed into products by processes that run indefinitely.

For some of these products, the concept of product design does not apply. But for others, product design is a critical task. Such products are known as *formulated products,* and are made according to a recipe, or formula. The design challenge is to identify the formula that enables the product to best meet potentially conflicting functional, processing, and cost requirements. This chapter focuses on design issues unique to formulated products and the ways that knowledge-based systems help meet these needs.

Formulated products include plastics, rubber, glass, basic metals and alloys, coatings, adhesives, petroleum and its derivatives, pharmaceuticals, and chemicals. Some of these are sold to the ultimate customer in bulk form; others are made into discrete parts. Whereas discrete parts are distinguished by geometry and features, formulated products are distinguished by their chemical and physical properties.

The properties of formulated products are determined by the ratios in which ingredients are combined and the processing that is done to these ingredients. The design of discrete parts involves discrete choices. The design of formulated products involves continuous choices—the proportion of a given ingredient can be anything from 0 to 100 percent.

Manufacturers of formulated products face the usual competitive challenges of a worldwide economy. In addition, many formulated products are directly affected by the presence of new materials and new environmental requirements. Some raw materials are now regulated and therefore scarce or unavailable. For example, manufacturers of coatings are challenged to provide paint and other finishes that are durable, attractive, and easily applied, yet contain little or no volatile organic compounds (VOCs), a contributor to smog. Similarly, manufacturers of friction materials for brakes and clutches must find substitutes for the excellent thermal and mechanical properties of asbestos.

New opportunities exist as well. The commercialization of new materials such as high performance plastics, composites, and ceramics provides new performance and product opportunities. One design challenge is to achieve new properties, and another is to reduce costs for high-performance but high-priced materials.

Formulation can be a challenging task because in many cases the ingredients interact both chemically and physically. The nature of these interactions may be known in an anecdotal sense, but rarely can they be completely described in a theoretical or empirical sense. The number of ingredients and the complexity of the interactions can present a challenging design problem.

This often means that in the absence of sophisticated analytical tools, the best that can be expected is that a chemist expert in product formulation can identify the general direction in which a recipe should be changed to meet new requirements. The reformulation magnitude and details can only be established by experimentation.

Product properties may also be affected by processing methods. Again, the interactions are rarely fully understood. The need for better properties, lower costs, and perhaps new materials is causing producers to view the formulation task in a broader sense. Now, both recipes and processing must be optimized to best meet objectives of cost and performance. New materials, new constraints, and increasing competitive pressures make the formulation task more important and more difficult. These factors point to the need for better analytical tools and less reliance on trial-and-error.

6.2 THE NEED: A FORMULATION MODEL

What is needed is a way to explicitly relate formulations to properties so that designers can find the changes in formulations to provide needed changes in properties. What is needed is a formulation model that moves beyond anecdotal evidence to incorporate mathematical relationships based on either first principles or analysis of experimental data. This model would minimize the need for experimentation because the user could predict the properties that would result from a hypothetical formulation.

The need to reduce experimentation, and thus the need for a formulation model, is well known. Sophisticated, large organizations have implemented models based on statistics, and to a lesser degree first principles, for years. What is still needed are ways to move beyond these approaches and to enable smaller organizations without access to statistics or computational chemistry experts to achieve rapid, effective product design.

The challenge is to develop a model accurate enough to be useful at reasonable cost and with available resources. The number of ingredients, and complex chemical and physical interactions between them, generally preclude development of a useful formulation model based on first principles alone. For this reason, formulation models are typically based on relationships observed in data.

Issues associated with data-based models were reviewed in Chapter 4, as were the traditional methods of statistics and the knowledge-based approach of neural nets. The strengths and limitations of each approach were described in that chapter. The following discussion on modeling pertains specifically to formulation support. The issues of data quality and model validation are as important in formulation as in any application. The goal of minimizing experimentation will be reached only if the model is robust. The model must capture the subtleties in a given problem, including complex functional relationships, as well as abrupt changes in system behavior in different regions of design space.

A good model is a useful design tool because it provides accurate answers to the *what-if* question, the question of "What will happen to properties for a particular change in formulation?" But from the formulator's perspective, this what if capability may be necessary but not sufficient. To paraphrase Tyson Gill, former senior chemist at the Glidden Company, the goal is not to find out what if a particular formulation is tried. Instead, designers are interested more specifically in *how to get* new product properties.

The designer's interest is in meeting new objectives. These objectives may include performance, price, and manufacturability, and these objectives may be conflicting. The formulator may call upon a formulation model to do this. But since the model answers what-if, not how-to-get, questions, the formulator must guide the model through a directed search of what-if exploration. By intelligently re-asking the what-if question, the designer eventually learns to ask the right question, the one that yields the answer of the desired product attributes, and thus answers the more difficult how-to-get question.

Learning how to get certain properties is the essence of most design tasks. In real-world design problems, some design goals are critical, others less so. Further, some design goals may require that an exact value be achieved, but other goals may be more loosely constrained. Perhaps the value must be held within a certain range, and there may or may not be preferred values within this range.

Design recommendations that do not recognize these types of constraints are seen by many formulators as not responsive to real-world needs and therefore of theoretical interest only. Solutions that do not take advantage of prioritization and allowed ranges of values are seen to be not optimal because they are artificially constrained. And the current competitive climate is intolerant of less-than-best solutions.

The upcoming sections describe the various ways to develop a formulation model. Each method has strengths and limitations. One point to consider in the selection of a method is its ability to meet the goals of a formulation model. The model must, of course, be as accurate and complete as possible. But it should also be judged on the basis of its ability to tell the designer *how to get* certain properties, not just *what if* certain ingredients are mixed.

6.3 FIRST PRINCIPLES APPROACHES

As described in Chapter 4, models based on first principles incorporate fundamental laws of physics or chemistry. The ability to trace a model back to fundamental principles provides a firm theoretical basis for these models. If this approach is feasible, it is preferable because it avoids reliance on data that themselves are subject to error.

The alternatives to first principles—statistics and now knowledge-based systems—are used because in most cases first principles are not sufficient. Too many simplifying assumptions must be made, and the model cannot be made sufficiently accurate. However, there are cases in which first principles are useful in the design of formulated products. A class of software products known as *molecular modeling* has emerged. These tools help designers identify and design molecules that achieve desired properties. They do so on the basis of first principles regarding the structure and behavior of molecules.

Tools such as HyperChem by Hyperchem Inc. incorporate 3-D graphics to enable users to visualize the structure of molecules. Molecules can be visually manipulated and combined to build larger molecules. Some properties of these molecules can be inferred from their structure, as established by the nature and the properties of the bonds linking individual atoms together into molecules and individual molecules into chains of molecules.

Molecular modeling also calls upon first principles commonly known as *computational chemistry*. HyperChem provides information such as energy, heat of formation, ionization potential, electron affinity, dipole moment, and ultraviolet, visible, and infrared spectra. Expert chemists can use this to predict properties of a molecule, and in this way find or *design* a molecule to meet their requirements.

Molecular modeling is useful in the following circumstances:

- Molecular properties can be adequately described by computational chemistry and molecular structure.

- An expert computational chemist is available who understands and can take advantage of these intrinsic properties.
- The properties of the molecule itself are of interest rather than, for example, processing of the molecule or the combination of various molecules in blends or composites.

Molecular modeling is a useful technique in design applications for which molecular structure is the sole design variable. Many formulation tasks, however, are broader in nature and thus cannot be solved in this way. For example, products consisting of mixtures derive properties on the basis of the interaction between ingredients. Thus, the primary design task is one of optimizing the proportion of ingredients, not the structure of any one ingredient. For such mixture problems, and for problems in which processing is a relevant design variable, the following modeling methods are used.

6.4 STATISTICAL APPROACHES

Many traditional formulation models are based on statistical methods. Some users build these models from scratch, programming statistical routines found in textbooks. The more common practice is to use commercial general-purpose statistics software.

The goal is to find equations that will serve as formulation models. The form of equations and the constants in these equations must be found. This is done by techniques such as **regression** analysis. A mathematical relationship between the variables is assumed, and on this basis, the corresponding form of equation chosen. Constants for that equation are then selected to minimize the differences between known data points and the corresponding points predicted by the model. This process may be viewed as *curve fitting*.

The equation used may have been selected on the basis of an understanding of underlying relationships in the formulation of a particular product. Alternatively, the user may simply try a number of equation forms, and a range of values for each of the constants, and from this select the equation and constants that provides the minimum total error between real and hypothetical data points.

Conceptually, this a straightforward process. If the user has no theoretical basis on which to select a form of equation, then a process of trial and error will lead to an acceptably good solution. In simple cases, this is an effective approach because the user can easily validate the results. In the case of single-variable systems, or when the variables can be meaningfully treated as independent variables, the user can produce meaningful two-dimensional graphs that, by visual inspection, indicate the accuracy with which the statistical model (as represented by a curve) matches reality (as represented by individual data points).

However, in most formulation problems, the ingredients are not independent. Their individual effect on properties is related to the relative proportion of more than one, if not all, of the other ingredients. A model that cannot represent multiple interactions will be limited in its ability to find the best way to meet new performance, cost, or quality requirements. If the model can vary only one parameter at a time, while holding the others constant, the best that can be done is to do many trials in which a single parameter is varied and try to uncover mutual interactions from these limited trials.

This simple, if not elegant, approach will lead to errors. The user must determine whether these errors are acceptably small. As with any modeling application, this is done by comparing predicted values to real values. These real values may be provided by experimentation or monitoring of production data.

Both statistical methods and neural nets presume to discover truth through data. This means that the data must tell the entire truth. In practice, this can be a tall order. In many applications, data are scarce or incomplete and contain errors.

Skilled users of statistics understand the implications of these situations and use a variety of data quality tests to analyze the distribution of data to identify data points that are suspiciously far from others and the range over which statistically significant data is available. The range and theoretical accuracy of a statistical model can only be established once the quantity and quality of available data are known.

An analysis called design of experiments (DOE) can be used to identify sparse regions in data space and to identify the experiments that should be performed to fill in these sparse regions. The goal is to acquire data that demonstrate the interaction of all parameters over the full range of all parameters. If n-dimensional data are viewed as being plotted in an n-dimensional space, DOE will identify the tests that should be run to provide sufficient data to populate all regions, including the vertices and the interior of an n-dimensional cube. In the absence of such complete data, there will be a measure of uncertainty regarding the relationship between parameters, and thus the validity of the model.

A variety of commercial general-purpose statistical analysis packages are available. Products such as RS/1, from BBN Software Products, and SAS System, from SAS Institute, Inc., provide a variety of statistical analysis and graphing capabilities. Common calculations include mean, median, and standard deviation (to determine the distribution of data points), correlation (the degree to which one variable affects another), t-test (to determine whether the difference between the means of two data samples could have occurred by chance), and the Kolmogorov-Smirnov test, (to determine whether a given parameter came from a truly random source). Some packages also provide nonlinear regression analysis, for more accurate modeling of nonlinear relationships, as well as multivariate-regression analysis, to support investigation of the simultaneous impact of multiple parameters. These two capabilities are useful in many formulation applications.

Some statistical packages incorporate traditional user interfaces consisting of lines of text, perhaps enhanced by simple graphs or forms. More current products employ so-called **graphical user interfaces (GUIs).** GUIs represent a radical transformation of the look and function of the user interface. Screens consist of one or more *windows,* work areas that approximate the look and function of individual pieces of paper on a desk, each dealing with a particular task. Some GUIs represent programs or tasks pictorially through the use of icons. As an example, the Apple MacIntosh GUI, often considered the benchmark, uses a wastebasket icon to represent file deletion. A file is placed in the wastebasket to throw it away.

Many commercial statistics packages are sold as modules, with each module offering a particular analytical or graphing capability. Pricing ranges from hundreds of dollars for PC-based modules to thousands of dollars for workstation-based modules, to tens of thousands of dollars for multiuser licenses on powerful workstations.

Statistics-based tools are well known, and therefore a comfortable approach for many users. They have been successfully used for formulation development. Nonetheless, there are limitations evident in many formulation applications. This has generated interest in alternative approaches.

6.5 KNOWLEDGE-BASED APPROACHES

Knowledge-based system technologies support formulated products in several ways. Some design support tools combine expert systems and materials databases. The expert system rules help the user select materials to meet specific functional requirements. This approach can be compared to the use of handbooks and other references by an expert chemist. Neural nets are also used for product formulation. They can be compared to the most common competing approach—statistics. In both cases, benefits can be attributed to the use of knowledge-based technologies.

6.5.1 Expert Systems

Expert systems contain knowledge in the form of rules regarding the analysis of a problem or the completion of a task. In the case of formulated product design, an expert system would include rules regarding some aspect of product design. These rules might call upon data within the expert system or from an external source. In use, the system would perform in the same manner as a human expert, typically a chemist, calling upon properties data and design rules in handbooks or other references.

An example of such a system is EXPOD by Mitsubishi Research, Inc. This workstation-based system is designed to support the development of polymers, in particular linear homopolymers and alternating copolymers. EXPOD was developed in collaboration with Showa Denko K. K., Sumitomo Chemical Co., Ltd., Toray Industries Inc., and Mitsubishi Kasei.

The joint research and expert system development effort produced a database of over 4,500 molecular structures and 3,100 known polymers, and a knowledge base that contains 59 rules for predicting properties and 130 facts regarding chemical structures and properties. A graphical user interface provides access to the database and knowledge base and provides graphical representation of molecular structure. These data and knowledge are used to estimate 13 properties such as density, dielectric and thermal expansion coefficients, and characteristic temperatures, all of which are intrinsic to a particular molecular structure. This what-if exploration is accomplished by specifying a desired polymer in the database and selecting an estimation method from the knowledge base. Users may also improve estimation accuracy for polymers of interest by incorporating their own experimental data.

Another EXPOD module performs *reverse inference*. Rather than estimating properties for a user-specified polymer, this module enables the system to search its database to find candidate polymer structures meeting user specified requirements. These requirements may include polymer classes such as polycarbonates and initial forms such as bisphenol-A polycarbonate. Upper or lower limits on properties may also be specified. The how-to-get results will be diagrams of polymer structures that are estimated to meet the specified property limitations.

This expert system shares strengths and limitations of other expert systems. Its strengths include valuable information in the form of data and expert knowledge. For the proper applications, this can provide great savings in experimentation and staffing. The limitations, of course, are imposed by the bounds of that data and expertise. In this case, the absence of data on blends and composites as well as random, block, and graft copolymers is a limitation. Although the system is designed to incorporate user-supplied data and knowledge, the system will be of benefit primarily to those whose applications fit within those for which the system was designed and for which data and knowledge were provided.

The need for more general-purpose tools is addressed by statistics, and now neural net-based systems, as described below.

6.5.2 Neural Nets

Neural nets find relationships in data. These relationships may be seen by using unsupervised learning to determine that formulation data can be meaningfully grouped, or clustered, according to the value of one or more parameters. In many cases, it is possible to draw useful conclusions on the basis of clustering. For example, different data clusters may represent different but equally effective ways to meet certain requirements. Or, the data as a whole may represent a general design problem such as tire design, and each cluster may represent a particular trade-off between conflicting properties such as tread life and traction. Observing such clusters can provide the designer a good starting point for a new design. To design a tire with excellent tread life, the designer would start with a formulation similar to those in the *maximum tread life* cluster.

Neural net-based supervised learning may also be used to identify a repeatable mathematical relationship between product formulation and product properties—that is, a formulation model. This model is used to avoid needless experimentation by estimating (rather than guessing) the properties for a new formulation. Quite a few of the larger formulated-products companies now use neural nets for this purpose.

Product formulators such as Tyson Gill and Dr. Joel Shutt have experience in the application of both statistics and neural nets for the development of formulation models for the Glidden Company and Lord Corporation, respectively. They note that statistical methods are widely used and a comfortable approach for many. They also note that neural nets offer the following benefits:

- Neural nets are inherently nonlinear and of unlimited order: The formulation model is not artificially constrained to be linear or of less than, for example, second order.
- Neural nets are more tolerant of noisy data—that is, data with gaps and errors.
- Neural nets are better able to extract information from data not meeting formal design-of-experiment criteria, a common situation due to limitations in the time and facilities available to run experiments.
- Neural nets can handle both discrete and continuous variables simultaneously, unlike statistical methods, which are designed for one variable type or the other.
- Neural net-based models can incorporate both ingredients and process parameters, a challenging task with statistical methods.

The above benefits were identified in formulation applications in coatings and adhesives. Similar results have been reported in the formulation of rubber, plastics, specialty polymer products, friction materials, metal alloys, pharmaceuticals, and biotechnology products. These findings are interesting because they add a dimension to a discussion of the merits of knowledge-based technologies. Users report benefits that go beyond the acquisition of knowledge into new and quantitatively better ways to use knowledge.

The neural net software that is used to develop formulation models comes from a variety of sources. Some users have developed neural net-based systems from scratch, based on equations available in the literature, or from **shareware,** free (but often unsupported) software available on computer bulletin boards. These users are generally very knowledgeable about the nature and application of advanced computer technologies.

Some of these users find, in the course of their first implementation, that the limitations in this low-cost approach, including limited capability and sophistication of the software and the absence of documentation and support, are such that this is not actually a cost-effective approach when the demands on the developers are considered.

Commercial neural net development system software meets the needs of users who find their time to be better spent solving formulation and other application-specific problems rather than functioning essentially as software developers. Commercial products offer a variety of neural net architectures. PC-based products start around $100, and most workstation-based products are in the range of thousands of dollars. The more expensive packages offer enhanced capabilities based on more sophisticated software design, more sophisticated user interfaces and documentation, and designed-in capabilities to link to other software for embedded applications.

The products reflect a diversity of neural net strategies. Neural Ware and others offer a wide variety of neural net architectures to meet different needs and preferences. An alternative approach is provided by AIWARE's Functional Link Net, which provides a single architecture that supports both unsupervised and supervised learning tasks, and thus provides an integrated environment for both.

Even with documentation that ranges from minimal to excellent, these tools require a fair amount of interest on the part of the user in the details of neural network application. A neural net architecture must be selected along with tuning parameters. These tuning parameters affect the rate at which the neural net converges on a solution and the accuracy of that solution. Without expert assistance, many users find this step to be mysterious or at least time-consuming. Suitable training parameters can be found by trial and error or by consultation with experienced users.

The use of a neural net also requires pre- and postprocessing tasks similar to those required for statistical methods. Preprocessing tasks include data acquisition, judgment of the quality and completeness of the data, and repair of the data as required. Repair tasks include filling in gaps within individual records and adding records as required to cover the intended range of the model. The user must also identify and resolve conflicting data, such as data in which a given effect is associated with two different causes. The user may also need to test whether any parameters are random, thus adding noise to the model.

Postprocessing tasks include the obvious tasks such as presentation and interpretation, using a spreadsheet or graphical format. The larger postprocessing task for formulation is to identify an optimal formulation, answering the how-to-get question. This may be done by repeatedly asking what if or by use of the method described in the next section.

6.6 AN INTEGRATED TECHNOLOGY DESIGN APPROACH

AIWARE's experience in formulation modeling, beginning in the mid 1980s, indicated that these applications shared common requirements and that an opportunity existed to meet these requirements with a neural net-based system designed specifically for this application area. It was believed that a product opportunity existed if an automated how-to-get capability could be provided.

In the course of developing these several formulation support systems, AIWARE discovered the difficulty in answering the how-to-get question. The obvious solution of developing an inverse formulation model was tried. Instead of starting with formulations, and from these predicting properties, the idea was to start with properties and then estimate formulations.

Unfortunately, this approach did not work. The formulation model was neural net-based, but the limitations found were not due to neural net capabilities. The larger issue was that if a mathematical mapping (or model) was found from formulation to properties, a unique mapping did not exist in the reverse direction. The inverse model would estimate a formulation that, at least theoretically, would provide the given properties. But this formulation was generally not the one that would best meet the multiple objectives of cost and performance while accommodating processing and materials constraints.

It was discovered that the real need was not for a model alone but for a model integrated with an optimizer. In this way, how-to-get questions were directly answered.

The following section describes the application of a product called CAD/Chem (a computer-aided design tool for chemists). This product is one of the first to integrate knowledge-based technologies, and it is one of the first knowledge-based products to focus on a specific application. The decision to combine technologies was made for the following reasons:

- Neural nets were effective for formulation modeling but only provided what-if answers.
- Users needed how-to-get answers that reflected constraints and priorities.
- Users wanted to focus on formulation tasks and not be distracted by neural net or other KBS details.

The objective of an easy-to-use how-to-get capability was achieved with a neural net-based formulation modeler, an integrated optimizer, and an embedded expert system functioning as a system manager. This combination of technologies solves how-to-get formulation problems in the following way:

- Neural net models estimate properties for a given formulation.
- An expert system runs these models repeatedly, in essence asking many what-if questions.
- Users define design goals, using a fuzzy logic-inspired method to specify ranking and preferences for product

properties and constraints in ingredients, processing, and costs.
- An optimizer drives the repeated what-if trials in the direction required to meet these goals.

CAD/Chem setup and tuning are largely automated, so that it is seen as a chemist's design tool, not a knowledge-based system as such. Although it is a general-purpose modeler/optimizer, the design decision was made to keep it simple from the user's perspective. Thus, many of the terms used in describing functionality and application are specific to product formulation rather than modeling or knowledge-based systems in a more general sense.

The first step in the use of CAD/Chem is to develop a neural net-based formulation model. The user begins by importing data in the common form of a delimited ASCII file. This file contains formulated product records, each of which contains proportions or absolute quantities of ingredients, process settings if applicable, and product properties. The arrangement of records within the file is not important because the user will specify as part of the **load data** operation which numbers represent model inputs (ingredients and process settings) and which are outputs (resulting properties).

The **data analysis** function is then used to clean up data by identifying duplicate or conflicting records and unchanging or linearly dependent variables. Conflicting records are those in which one set of inputs is shown with more than one set of outputs. This situation is not uncommon and may be caused by experimental or data-logging errors. But since conflicting records can destroy a model, they must be resolved. The user may eliminate or assign lower belief values to the less-credible conflicting records. In a similar vein, linearly dependent or unchanging variables create unnecessary work for the model, and so they are typically eliminated before neural net training.

An analysis of correlation between ingredients and properties is provided as a way to determine the relative impact of various ingredients. This correlation analysis, determined by a nonparametric statistical method, is shown in Figure 6–1. In this case, the positive or negative correlations of proportions of ingredients, such as silica, to the property of cure time are displayed by the appropriate ingredient. The user may choose to eliminate parameters

FIGURE 6–1
CAD/Chem Correlation Screen

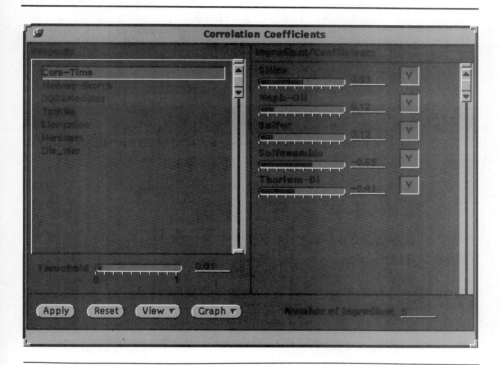

below a chosen correlation threshold for a particular CAD/Chem run to verify whether that ingredient is required. If the ingredient is not required, processing can be simplified by removal of the extraneous ingredient.

The user then **trains** the system, that is, develops the neural net-based formulation model. Upon completion of training, the functions of **best match, estimate,** and **optimize** are available through the **consult** screen shown in Figure 6–2. Rubber compound ingredients are shown in the Given column. The Estimate Property function, initiated by the pushbutton at the bottom of the screen, has provided the product properties shown in the Found column on

FIGURE 6–2
CAD/Chem Consult Screen

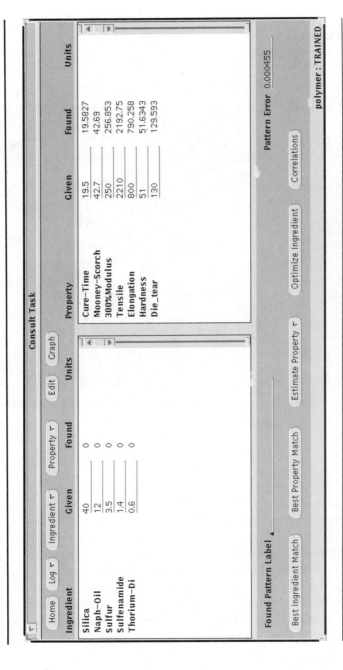

the right side of the screen. This Estimate Property is the what-if functionality and provides the answer to the question "What properties will result if ingredients are combined in the proportions shown on the left half of the screen?"

In this case, estimated properties displayed in the Found column are used to perform the important task of model verification, that is, assessment of model accuracy. This is done by asking the system to estimate properties for which the correct answer is already known. The ingredients shown represent a recipe for which experimental property data exist. These experimental data are displayed in the Given column. The system predicts properties for the same recipe, and these estimated properties are shown in the Found column. Comparison of the numbers in the Given and Found columns show close agreement. Thus, the underlying model may be assumed to be accurate, at least in the region of the recipe shown.

Pressing the Best Ingredient Match or Best Property Match pushbutton shown in Figure 6–2 will cause CAD/Chem to prompt the user to specify a set of ingredients or properties, respectively, and then provide the user the prior formulation that provides the closest match to the specified list of ingredients or properties. If the user specifies some but not all ingredient values, CAD/Chem will identify the prior formulation providing the best match to the specified ingredients, without any constraints placed on the unspecified ingredients.

In this case, the user has chosen to use the properties in the Given column and found in the database with the label "test-5" seen at the bottom left of the screen attached to the record. Best Match may be useful to find a starting point for a new design. This reference to prior formulations may also provide access to other useful information in a staff member's memory or organizational records.

Users report that the most useful functionality is *optimize*. This how-to-get analysis tells the user how to best meet multiple objectives. These objectives may include constraints on levels of ingredients. The user may specify these constraints using the screen shown in Figure 6–3. Since CAD/Chem treats process conditions in the same way as ingredients, both may be incorporated in the formulation model, and constraints may be specified for both. In this case,

FIGURE 6–3
CAD/Chem Constraints Screen

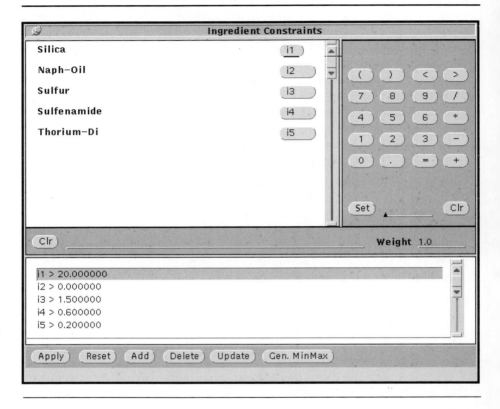

the user has specified that ingredient 1 (silica) must be at least 20 units, ingredient 3 (sulfur) at least 1.5, and so on.

Turning to property objectives, the Property Weights screen in Figure 6–4 is used to specify the relative importance of properties. In this example, Cure Time is shown here as most important, and Hardness as least.

As the optimizer seeks to satisfy multiple objectives, less importance is assigned to properties with lower weights, and thus less important properties are not allowed disproportionate effect when trade-offs must be made. By trying different property weights, the user may also perform a sensitivity analysis.

FIGURE 6–4
CAD/Chem Weights Screen

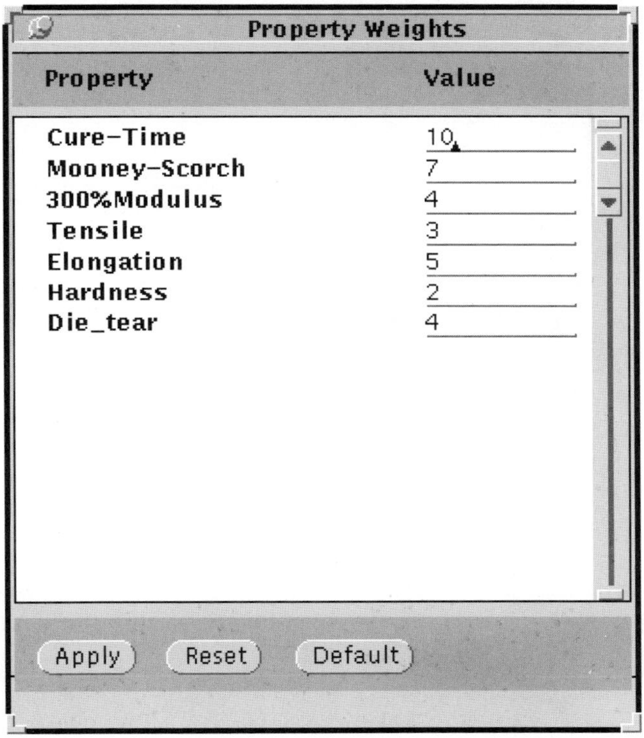

Property Weights	
Property	**Value**
Cure–Time	10
Mooney–Scorch	7
300%Modulus	4
Tensile	3
Elongation	5
Hardness	2
Die_tear	4

Apply   Reset   Default

The user may further define property preferences using the Desirability Function screen shown in Figure 6–5. Desirability functions enable the user to specify acceptable ranges for each product property and to specify preferences within the specified range. The ability to specify ranges and preferences within the ranges is critical to achieving optimal formulations. In most cases, trade-offs must be made between various properties and cost. It would be unduly restrictive to hold a given parameter to a single value when a value held anywhere within a range would be acceptable. By providing this design latitude, the user is able to achieve desired properties in the least expensive way. The user may try several specifications

FIGURE 6-5
CAD/Chem Desirability Function Screen

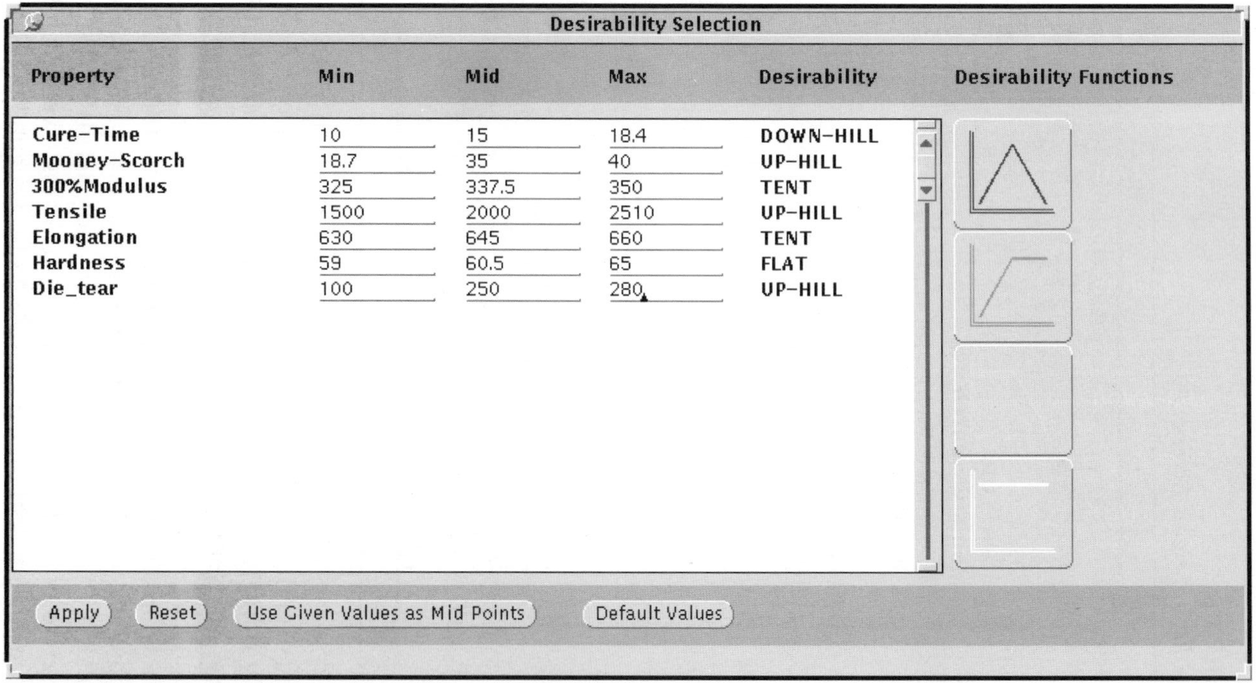

Desirability Selection

Property	Min	Mid	Max	Desirability
Cure–Time	10	15	18.4	DOWN–HILL
Mooney–Scorch	18.7	35	40	UP–HILL
300%Modulus	325	337.5	350	TENT
Tensile	1500	2000	2510	UP–HILL
Elongation	630	645	660	TENT
Hardness	59	60.5	65	FLAT
Die_tear	100	250	280	UP–HILL

Desirability Functions

Apply Reset Use Given Values as Mid Points Default Values of desirability functions and costs to explore cost versus performance trade-offs.

The four Desirability Functions shown on the right of the screen offer four ways that preferred values may be distributed within the range of acceptable values. In each function, the horizontal axis represents the parameter value, and the vertical axis represents relative desirability, with desirability increasing with increasing distance above the horizontal axis.

The chart to the left of the functions is filled in by the user to specify the Min(imum) and Max(imum) values that define the range of acceptable values. The Mid point indicates the break point in the Desirability Function chosen. For example, in the TENT icon, the Mid point indicates the peak—that is, the most desirable value of that parameter. The UP-HILL function is used to express increasing preference as a parameter progresses from the given Min to the Mid point, and equal satisfaction for that parameter value anywhere between the Mid and the Max point. The FLAT icon indicates the user finds all values within the specified range to be equally desirable.

The user may now initiate the optimization process, using the screen shown in Figure 6–6. Ingredients and process parameters, and initial values of each, are shown on the left side of the screen. After pressing the Optimize button at the bottom of the screen, the user will observe property values changing every second or so as CAD/Chem runs a succession of what-if studies.

After each what-if, the optimizer determines how well the specified multiple objectives have been satisfied. The degree to which individual property objectives have been met is displayed by the thermometer graphs next to each property, and the degree to which overall objectives have been met is shown by the Total Desirability thermometer at the top of the screen. The optimizer then defines the next what-if study to be performed and repeats the process.

Initially, the property values jump around, but as the optimizer closes in on the best solution, the property values stabilize. Total Desirability will climb to the highest possible value, as will the individual Desirabilities.

Ideally, all Desirabilities will end up at 100 percent. In practice, this may not be possible. The best feasible solution will be identified for the circumstances defined by the model and the specified constraints and preferences.

FIGURE 6–6
CAD/Chem Optimization Screen

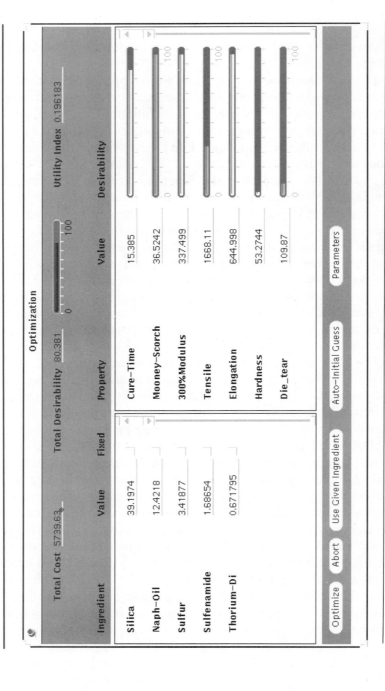

Since every property has a price and every process its limitations, users may want to perform a sensitivity analysis. The user may choose, for example, to explore the sensitivity of a critical property to variations in the levels of a particularly expensive ingredient. Compensating changes in formulation using less expensive ingredients may also be identified.

If the user has incorporated process parameters in the analysis, the sensitivity of critical properties to variations in process conditions may be explored. The user may find that current process capabilities are insufficient to ensure that target product properties will be met and so a nominally less desirable but less sensitive formulation is required unless better control capabilities can be achieved. Having determined that the cost of better control of parameter variations will be repaid by improved product properties or reduced costs, the user may also use this information to justify better control capabilities.

This analysis is supported by the 3-D graphing capability shown in Figure 6–7. Ingredients, process conditions, properties, and desirabilities may be assigned to the three axes as the user wishes. The most common approach is to assign to the vertical axis either the value of a particular property or the ability to meet that property requirement—that is, the desirability achieved for that property. In the example shown, the vertical axis indicates a property, Mooney scorch. The horizontal axis represent proportions of two ingredients, sulfur and silicon. The shape of the response surface provides an understanding of relationships that would be difficult to achieve with numerical output alone. The user may also *click on* points on the response surface to establish the exact values of the three coordinates in regions of interest.

6.7 APPLICATIONS AND PAYBACK

Applications in coatings, plastics, and rubber formulation were instrumental in designing and validating CAD/Chem. Further applications have been reported in specialty polymers, building materials—including floor and ceiling coverings, integrated circuit fabrication, pharmaceuticals and biotechnology products, automotive and truck tires, friction materials, and a variety of chemicals. Users

FIGURE 6–7
CAD/Chem 3-D Graphics Screen

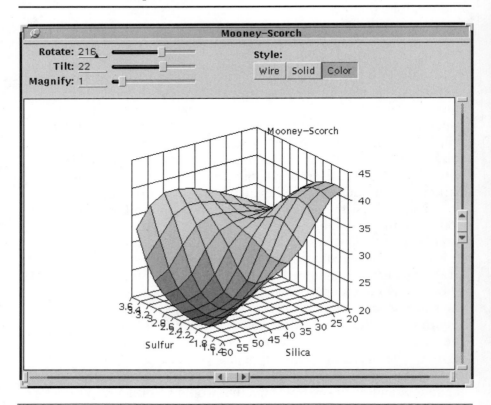

report similar benefits. Quantifying these benefits is a challenge since many users of knowledge-based systems consider these systems to be a strategic advantage and therefore proprietary. What little information is available enables the following conclusions.

6.7.1 Equipment Housings

Plastic used in electronic equipment housings was to be reformulated to provide electrical conductivity, so that cabinets of this material would function as built-in electromagnetic interference

(EMI) shields. This would avoid the need for a traditional EMI shield consisting of a separate metal mesh enclosure.

CAD/Chem identified an acceptable formulation in approximately two hours, including data loading, system setup, and product optimization. The client indicated that a similar design task had previously required three months of effort by their formulating chemist and lab technician.

Ignoring savings in facilities costs and benefits from faster time to market, the following rough estimate of payback can be made on the basis of salary and benefits savings alone. The three calendar months saved are assumed to correspond to a total of three person-months between a chemist and a technician. A salary of $40,000, and benefits of 30 percent are assumed to represent an average between the two individuals.

The savings from the first use of CAD/Chem would be (3 mo./ 12 mo.) $\times$ ($\$40,000 \times 1.3$) = $\$13,000$. The payback period would depend upon required investment. If a $\$10,000$ software license were to be purchased, the payback period would be

$$\frac{\$10,000 \times 12 \text{ mo.}}{\$40,000 \times 1.3} = \text{approximately 3 months}$$

If a workstation, peripherals, and multiuser license were purchased for $\$35,000$, the payback period would be

$$\frac{\$35,000 \times 12 \text{ mo.}}{\$40,000 \times 1.3} = \text{approximately 8 months}$$

6.7.2 Structural Adhesives

A structural adhesive was reformulated by the Lord Corporation in response to a new market requirement. To simplify application of the adhesive, the client suggested that the properties of a two-part adhesive be achieved with a single-part adhesive. As reported by Dr. Joel Shutt, this complex design task was pursued for one year without success. Upon acquisition of CAD/Chem, a formulation was identified in days.

In the absence of a comparable design task, the time to complete this task without the assistance of CAD/Chem is unknown. However, it is estimated that perhaps one more year would be required. What might have been a two-year development program was reduced to days.

The payback period for this application would be similar to that of the previous examples, depending on salary and benefits costs. With costs assumed to be similar, the payback could again be anticipated in three to eight months.

6.7.3 Coatings

A coating reformulation project was undertaken by the Glidden Company, with the goal of improving the coating's resistance to contact with acid. As reported by Mr. Tyson Gill, a series of experiments was undertaken and test data fed to CAD/Chem. The optimizer generated a 3-D graph of monomer and Tg levels versus acid resistance. The resulting graph looked like a U, with highest (best) acid resistance corresponding to the tops of the U-shaped response surface.

Inspection of this surface showed that good acid resistance was possible with both high and low levels of monomer. At first, the latter result was dismissed, since conventional wisdom held that relatively large amounts of the expensive monomer were required for good acid resistance.

Upon further investigation, it was revealed that there had been other occurrences of good acid resistance with low monomer levels. Jogging the organization's collective memory in this way provided a significant cost savings by lessening by 96 percent the proportion of this expensive monomer in the product.

These applications illustrate the ability of knowledge-based systems to capture valuable knowledge for reuse and to identify critical product design relationships. These capabilities certainly provide long-term benefits, but in these examples, short-term benefits are also seen.

This chapter concludes the review of design applications. The upcoming chapters focus on knowledge-based systems for manufacturing, beginning with issues and applications in process planning and diagnostics followed by knowledge-based process optimization.

6.8 REFERENCES

1. CAD/Chem Custom Formulation System Technical Overview. Cleveland, OH: AIWARE Inc, 1992.

2. EXPOD product literature. Tokyo: Mitsubishi Research Institute, Inc. Menlo Park, CA: Mitsubishi Kasei America, Inc.

3. Gauthier, Michelle M. "Sorting Out Structural (Adhesives)." *Advanced Materials and Processes,* July 1990, pp. 26–35.

4. Gill, Tyson, and Joel Shutt. "Optimizing Product Formulations Using Neural Networks." *Scientific Computing & Automation,* September 1992, pp. 19–26.

5. Grant, Eugene, and Richard S. Leavenworth. *Statistical Quality Control,* 4th ed. New York: McGraw-Hill Book Company, 1972.

6. Hamburg, Morris. *Statistical Analysis for Decision Making.* New York: Harcourt, Brace & World, Inc., 1970.

7. HyperChem product literature. Waterloo, Ontario, Canada: Hypercube Inc., 1993.

8. Inglesby, Tom. "Industry Profile: An Interview with Larry Ford." *Manufacturing Systems,* October 1992, pp. 34–38.

9. RS/1 product literature. Cambridge, MA: BBN Software Products.

10. SAS System product literature. Cary, NC: SAS Institute, Inc.

11. VerDuin, William H. "Neural Nets for Custom Formulation." *Handbook of Expert Systems in Manufacturing.* New York: McGraw-Hill, Inc., 1991, pp. 515–24.

12. VerDuin, William H. "The Role of Integrated AI Technologies in Design and Manufacturing," *The Journal of the Adhesive and Sealant Council,* XXII, no. 1 (Fall 1992), pp. 269–77.

Chapter Seven

Knowledge-Based Process Monitoring and Diagnostics:
Improving Scheduling, Reducing Downtime

7.1 INTRODUCTION

In previous chapters, the focus was on the design of products and processes. In this chapter and the next, the focus shifts to manufacturing. The roles of knowledge-based systems are described as elements of a very different environment, with different needs and constraints and different skills and knowledge.

Common to all manufacturing is the need to provide ongoing and real-time support, monitoring processes to verify that all is in order and diagnosing problems when they arise. These tasks apply to discrete part manufacturing as well as batch and continuous production. This chapter focuses on the monitoring and diagnostic tasks and some knowledge-based approaches.

Chapter 8 covers knowledge-based approaches for the many manufacturing processes that can be optimized, or fine-tuned, to better meet objectives such as quality and cost. In both chapters, the need for rapid response and enlightened observation provides the motivation for better use of knowledge, and thus knowledge-based systems. For this reason, manufacturing support has been one of the dominant application areas for knowledge-based systems.

Returning to the subject of process monitoring and diagnosis, the task is to keep things running. A product and an associated process have been designed. The initial start-up bugs have been identified and resolved. Now, the process is to be commissioned and the product is to be produced in quantity, day in and day out.

This calls for a focus and skill set different from those of design. The design task calls for creativity and analytical skills. In production, problem-solving skills are critical. Immediate action is required, but the appropriate action may not be obvious. Fact-finding and analytical skills are required to uncover root problems that may be masked by nuisance alarms, irrelevant details, and other distractions.

The need for effective problem-solving skills and tools exists for ongoing process monitoring, but also for commissioning of a new process. Problem-solving skills are called upon to handle real-time surprises that may fly in the face of careful design and analysis. Experienced engineers are typically employed for process commissioning. They bring with them technical and problem-solving skills acquired from training and experience.

Once the production line has met acceptance specifications, the process and control engineers who got things running are often moved on to new assignments. The machine operators and maintenance staff inherit first-line responsibility for the tasks of ongoing maintenance, monitoring, and diagnosis. They handle the day-to-day challenges of equipment maintenance and repair—ideally scheduled, often not. They also contend with more subtle problems. Control bugs appear. Processes change slowly: Equipment ages, material characteristics change, and so conformance to product specifications suffers.

It would be great to have the process and control engineers who set up and thus really understand the line on call to diagnose if not fix these problems. This is often not possible. This raises issues related to both technology and people. New process technologies are implemented to provide better quality at lower cost. These improvements come at a price: The equipment is more complicated, the controls more integrated. As a result, the monitoring and diagnosis task is more difficult.

Integration helps make it possible for the impact of one problem to be reflected in, say, 50 parameters, each of which exceeds acceptable limits and thus sets off its own alarm. It takes advanced problem-solving skills, not to mention a tolerance for distraction, to uncover the root cause of the onslaught of alarms.

The operators and maintenance staff called upon to support the line will have the benefit of first-hand familiarity with the process and the equipment and the ability to see day-to-day variations in the product. But these staff members may be limited by their absence of training in problem-solving skills and their lack of familiarity with the advanced technologies underlying integrated processes and sophisticated controls.

The need for better monitoring and diagnosis skills is increasing, even as manufacturers find it increasingly difficult to find sufficiently skilled shop floor employees. This has led manufacturers to implement in-house training but has also provided the rationale for more comprehensive computer-based monitoring and diagnosis support tools. The capabilities of knowledge-based systems are well suited to these needs, as the following applications demonstrate.

7.2 EXPERT SYSTEMS APPLICATIONS

Expert sytems-based process monitoring and diagnosis systems are the most common knowledge-based system applications. The structured decision-making capabilities of expert systems mimic the logical troubleshooting approach of expert operators, maintenance staff, and engineers. These systems incorporate rule-based expert knowledge. They also call upon knowledge of current conditions.

This knowledge may be provided by the operator or maintenance staff person as part of the question-and-answer dialog between the expert system and the user to define the problem and identify a solution. This knowledge may also be provided by sensors, often the same sensors that provide input to the process controller. For purposes of control and diagnosis, accurate sensor data are critical. The question is, can a given sensor reading be believed?

7.2.1 Sensor Validation

Expert systems are used to validate sensors, that is, to determine if a given reading is credible. This is an important capability because a critical element in process monitoring and control is access to reliable data. This point receives a good deal of attention because of the risks involved. The risk posed by a bad sensor is that a process could fail, perhaps catastrophically, if control actions are based on a bad sensor reading. A monitoring and diagnostic system could also fail to identify a critical deviation if erroneous sensor data are used.

The importance of reliable sensor data would suggest the use of redundant sensors, that is, two sensors to measure one parameter. Two identical sensors could be used, and the presumably identical readings of the sensors compared. Or, to lessen the risk that both fail due to a single failure mode, one might measure a certain parameter in two ways. For example, the level of liquid in a process tank might be measured directly by a capacitive level sensor and indirectly by a pressure sensor at the bottom of the tank that weighs the liquid.

But to save installation costs, many processes forgo the benefits of redundant sensors and incorporate a single sensor per process parameter. This means that there is no explicit way to verify the reading of a given sensor.

In the absence of redundancy, other clues must be used to judge sensor validity. Sensors may fail in an obvious manner. If the sensor or its wiring are shorted out, this condition is obvious. If the sensor or its cable break so that the circuit is open, that condition is also easily detected. The insidious failures are those in which the sensor sticks, maintaining a plausible, but incorrect reading. A reading may have been correct at one time, but the unchanging reading would give no evidence of changes in the process. The control system reads what appears to be an acceptable reading and responds to what is actually an incorrect reading. The result may be degraded product quality or potentially catastrophic equipment failure.

Knowledge-based systems are used to validate sensor readings. Knowledge of process characteristics is used to judge whether a current sensor reading or an observed trend in that sensor's output is credible under the circumstances and, therefore, probably correct.

DuPont has reported wide experience in the use of expert systems for sensor validation. It finds this approach most useful for processes characterized by relatively slowly varying parameters, particularly those with dead times measured in minutes to hours. The expert system provides insight but can also provide the long-term monitoring that may be required to relate a particular reaction to a change that may have been made on the previous shift.

The expert system approach is less useful at the other end of the time spectrum. Rapidly changing systems may not give an expert system enough time to consider the validity of a sensor reading before the system moves to another state and the judgment process must begin anew. An example is high-speed motion control, in which the system can change state in milliseconds.

Sensor validation rules include common sense and knowledge about the process. For example, an operator would know that the temperature of a massive oven or reactor varies slowly. A rapid change in measured temperature would be suspicious because it is physically impossible. The expert system would track temperature readings and check that not only absolute values but also rates of change are within expected ranges.

Sensor readings of parameters such as flow and pressure may incorporate a characteristic level of noise in addition to the desired signal. An expert system monitoring noise levels might conclude that an increase in noise indicates an electrical malfunction. A decrease in noise would also be suspicious and may indicate that the sensor has become plugged or otherwise isolated from the system it is measuring.

Sensor data may also be validated by comparison to calculated values. For example, fluid flow may be roughly proportional to pump speed, and so pump speed may be used to provide a flow rate measurement reality test.

Process monitoring and diagnostic tasks typically require more than sensor validation alone. A system of larger scope is required to meet most manufacturing needs, and for this reason, sensor validation rules are often combined with process operation and troubleshooting knowledge.

The system begins by verifying sensor readings. If the readings are suspicious, they are flagged. If the sensors appear healthy, the system considers the health of the process itself. This is done by verifying that process conditions and product attributes fit within predetermined numerical constraints. If not, rules are followed to track down the cause. When a conclusion is reached, the system presents the user with a description of the problem, the solution, and if desired, the reasoning used to arrive at the conclusion. The use of knowledge-based technology in process monitoring provides the following benefits:

- Problem-solving expertise is captured before the holder of that expertise becomes unavailable due to reassignment or retirement.
- This expertise helps train less experienced staff.
- It provides consistent and high-quality diagnostics.
- The system can incorporate multiple perspectives and areas of expertise, for example, the machine operator's and maintenance staff's view of warning signs, symptoms, and effects of problems.
- The system can concentrate data, distilling volumes of data into more focused information.
- The system can either incorporate sophisticated analysis or provide automatic reference to that capability in an external package.

7.2.2 Forge Shop Scheduling

An expert system has improved overall forge shop operations by improving scheduling. Scheduling, experience, and logic are useful in balancing constraints such as process capabilities and inventory limitations and in meeting a variety of scheduling objectives. These objectives might include minimizing work in process, minimizing late orders, or minimizing machine setups. Expert systems have been shown to be useful in these tasks.

Ellwood City Forge (Ellwood City, Pennsylvania) specializes in open-die forgings of one quarter to 25 tons, generally with no two jobs alike. To address the scheduling challenges of such a business, the Carnegie Group (Pittsburgh, Pennsylvania) developed the Forge Shop Expert Scheduling System, incorporating company policies, metallurgy knowledge, and scheduling rules.

The system development goals were to achieve high quality and low cost with reduced order turnaround time and improved efficiency of material utilization. A secondary goal was to increase energy efficiency. The system had to manage up to 2,000 orders in the system, each of which could have unique materials and processing and many of which involve lot sizes of two or three pieces.

The following benefits were reported:

- Improved material yield and energy efficiency.
- More accurate estimates of price and delivery.

- Reduction of turnaround time from eight to less than three weeks.

These benefits were achieved through improved scheduling. The system schedules the melt shop, in which raw ingots, scrap steel, and alloying elements are combined. The system also schedules the forge, which implements the metalworking process that provides the shape and some of the physical properties of the finished part. This system also schedules the furnace that is used for the heat and quench processes that provide required levels of toughness and hardness in the forged parts.

7.2.3 Plastic Production Scheduling

Better scheduling capabilities have enabled pursuit of new business opportunities. In this application, a general-purpose PC-based scheduling expert system developed by Stone & Webster (Boston, Massachusetts) was applied by Union Carbide Chemical and Plastics (Charleston, West Virginia) to the task of finite scheduling, that is, determining a production schedule that best accommodates finite (limited) production capabilities. The development objective was to enable rapid, yet realistic, responses to opportunities that would periodically arise to rapidly produce products for export.

The decision on whether to accept an order had to be made within a day. If accepted, these new orders would be inserted into the existing production schedule, with some impact on the delivery dates of the existing orders. The operational challenge was to accommodate the impact of the added order without causing ship dates of existing orders to exceed prior commitments. The scheduling challenge was to rapidly determine whether this would be feasible before committing to the new order.

The scheduling system combines graphics, spreadsheeting, linear programming, and knowledge on ways to deal with demands that exceed capacity. When a constraint is violated by insertion of a new order, the expert system helps schedulers determine which constraints to override to meet demand. Schedulers may, for example, consult a screen that combines a Gantt chart and expert system results. The Gantt chart shows the time intervals and timing relationships for the various manufacturing steps. The expert system window displays scheduling rules and specific problem resolution advice to help users juggle conflicting demands and, hopefully, identify a feasible approach.

7.2.4 Welding

Welding is a popular expert system application area. Users perceive a need for specialized expertise in several related areas, including the design of welds as an attachment method, the welding process itself, and diagnosis and maintenance of welding equipment. In each case, the expertise is scarce and valuable. The value of capturing this expertise justifies development of expert systems.

The American Welding Institute (Knoxville, Tennessee) has developed a number of expert systems based on Texas Instrument's Personal Consultant Plus PC-based expert system shell. These systems help engineers to specify and produce high quality welds. WELDSELECTOR functions as a material/process advisor, helping the user select welding electrode materials for a given application. FERRITEPREDICTOR helps the engineer select a stainless steel welding filler. WELDSYMPLE provides designers weld specification symbols for use on mechanical drawings. WELDHEAT determines pre- and postheating requirements for optimum weld quality. WELDHARD helps specify weld cladding and hardfacing processes, while WELDSTRESS estimates the residual stress that will result from welding. SEAMTRACKER supports seam welding along a seam, a process that is made more complex by the need to accurately track the seam being welded.

Battelle Columbus Laboratories (Columbus, Ohio) has developed a diagnostic expert system called WELDEX that supports interpretation of the features found in weld radiographs. This knowledge covers identification of the size, shape, and location of internal weld flaws.

Stone & Webster has developed a WELDDEFECTS DIAGNOSIS SYSTEM to help managers identify the causes of weld defects. Their WELDER QUALIFICATION TEST SELECTION SYSTEM helps managers select the skill assessment tests to be used to qualify welders for specific welding tasks.

Deere & Company has developed a resistance welder diagnostic system to support both commissioning and maintenance of a new line of welders at the John Deere Harvester Works. This plant fabricates combines, machinery used by farmers to harvest grains. The introduction of a new line of combines was accompanied by introduction of a new line of resistance spotwelders to provide the 3,000 welds required per combine.

The CARS (Computer-Aided Resistance Spotwelding) System consists of approximately 130 spot welders, a host computer and network, cell controllers, and controllers for each spot welder. The spotwelder controllers are microprocessor-based.

Users program the weld schedules and fault parameters and monitor system operation with several LCD displays and 16 fault status LEDs. The controllers operate independently of the host computer if required, but the connection to the host computer enables quality and status monitoring and programming of the entire CARS System from a central point.

This system is significantly more sophisticated than the system it replaced. It was anticipated that the electricians, supported by maintenance engineers, would need training in the new system. In addition, the installation of the system was expected to place a heavy load on the already busy electricians. To address these issues, it was decided that a diagnostic expert system would be developed in parallel with implementation of the CARS System and that this system would be used to support training of electricians and commissioning and ongoing maintenance of the CARS System.

An expert system development team was assembled, consisting of an electrician, an electrical engineer, a maintenance engineer, and an electrical maintenance supervisor, as well as an engineer from the corporate production engineering department. The composition of this team provided a variety of skills and perspectives. The expert system rules were developed by consensus and thus fully used this range of input. Management support was reported to be critical in making team members' time available to sustain the development, validation, and implementation effort.

The system was designed to be used by qualified electricians. It was also assumed that a welder operator may not be available to support troubleshooting, as would occur during off-shift repairs. To cover the potential absence of this input during troubleshooting, the scope of the diagnostic system was extended to include fault indicators in peripheral pneumatic and chilled-water equipment, as well as indicators from the CARS System itself as inputs to the diagnostic system. For the same reasons, the diagnostic system provided, as the final diagnostic step, verification of all aspects of CARS System functionality.

Ease of use was a design goal. A user-friendly expert system shell was selected. This enabled the transfer of programming duties partway through the development effort from the corporate engineer to the plant engineer. The system was made easily portable by installation in a 286 laptop computer, which was mounted in a briefcase along with all of the drawings and documentation for the CARS System.

The knowledge acquisition phase of the project involved the usual challenges of ensuring continued access to the entire development team and maintaining motivation of all members. The electrician serving as one of the primary domain experts became fully supportive of the project only when he was convinced that the system would be used only by qualified electricians and that it would support, not replace, his peers.

Knowledge acquisition is an intensive task that, to be effective, requires clarity and focus. The development team found that knowledge acquisition sessions were most effective when limited to three or four hours. The team met biweekly, drawing information from in-house experts, CARS equipment vendors, and system documentation. A case study approach was employed. Hypothetical symptoms were defined, and problem analysis and solution approaches were identified by consensus.

System verification is a necessary element of any knowledge-based system development. In this case, verification was made more challenging by the fact that the diagnostic system was developed in parallel with installation and commissioning of the CARS System itself. For this reason, only part of the diagnostic system was verified by use on the real CARS System. The balance of verification was performed using simulated problems and symptoms.

This approach enabled the diagnostic system to achieve maximum utility as a training and start-up support tool. However, since the system was partially verified on simulated problems, its accuracy and completeness were somewhat diminished. System accuracy was reported to be 80 percent—that is, 80 percent of the problems presented to the system were correctly diagnosed and rapidly resolved. Electricians updated the 400-rule system as required when a correct diagnosis was not provided.

The system is considered a success. It is reported to be used on a regular basis and to reduce the frequency with which the experts must be called. The training provided by the system has provided flexibility in scheduling of electricians, since the system enables qualified electricians who are not expert on the CARS System to provide effective maintenance of this sophisticated new system.

System enhancements have been identified and initiated. An audit-trail capability is being installed to log the date and time of system use, the user's name, and any comments such as system effectiveness in resolving specific problems. A planned enhancement is to add a graphics capability to replace paper-based documentation, including mechanical drawings and electrical drawings of the equipment.

This application illustrates a task for which expert systems are well suited—proceeding through a structured analysis of symptoms to find what is usually the single root cause of a problem. This application also demonstrates effective management of people-related development issues.

7.2.5 Food Processing

People-related issues are also visible in one of the most famous expert system implementations, the Campbell Soup COOKER.

To provide the virtually unlimited shelf life of canned food, the food and the can are sterilized. This process is performed on a commercial scale by hydrostatic canned food product sterilizers that are 30 feet square, 70 feet high, and process 10,000 cans per hour.

In 1984, Campbell Soup Company made soup at six locations worldwide, and each location had at least one sterilizer. Expertise in the diagnosis and repair of these sterilizers resided primarily in a man with 44 years of experience, 14 as Campbell's roving sterilizer expert. His imminent retirement presented a dilemma, because his expertise was critical to minimizing expensive and disruptive downtime.

It was determined that the then-new expert systems technology would be used to capture and preserve this expertise. Repair delays imposed by the need for the expert to travel from plant to plant would also be minimized.

The system, named COOKER, was developed by Texas Instruments (TI) using its Personal Consultant expert system shell programmed in LISP, running under MS-DOS on a TI Personal Computer. COOKER was developed in seven months. With the addition of rotary cooker rules to those of the hydrostatic cookers, the basic system included 151 rules. Copies of COOKER, sent to each of the plants, were customized to incorporate rules regarding equipment start-up and shut-down unique to each location.

The system was extensively used, even before retirement of the expert. This success led to development of an additional seven systems from 1984 to 1986. Each supported diagnosis of a specific type of equipment used to form cans or process soup, as follows.

CLOSER captured the expertise of a manufacturing engineer with 51 years of experience. The five-month development of this system captured expertise on the machinery used to process cans fabricated of walls, top, and bottom as three separate pieces. COIL LINE was developed in four months, capturing the expertise to diagnose Littel Coil Lines, the equipment used to unroll coils of tin-plated steel and produce the flat panels from which can components are stamped in a separate process. FILLER was also developed in four months and captured the expertise required to diagnose problems in can-filling machines. FILLER and COIL LINE incorporated the expertise of one engineer.

The following systems also called upon the expertise of one engineer, supplemented with information from maintenance and operations guides. These four systems were developed between August 1985 and May 1986. SHEAR diagnosed problems with the scroll shear machine used to cut metal sheets from the Littel Coil Line into scrolled strips. 314-AP PRESS supported the press that cut can ends from scrolled strips. CURLER diagnosed the machine that forms the curl on the ends of the can wall that is then folded over the edges of the top and bottom pieces. CAN ASSEMBLY supported forming of the can body and sealing of the side seam and was the only expert system in this group to incorporate a graphical user interface.

These systems were reported to be widely used and to save Campbell Soup Company thousands of dollars per use. But for several reasons, Campbell no longer uses these systems.

Computers located near process equipment are mounted in waterproof and thus washable enclosures. When Campbell changed its standard operating system from MS-DOS to OS/2, new computers placed in these washable enclosures were OS/2-based. Continuing to use the expert systems that were developed under DOS would require that the software be redeveloped to run under OS/2 or that DOS machines be maintained along with the new OS/2 machines. The expense of watertight enclosures could not be justified for the older DOS computers. Thus, the DOS-based expert systems were relegated to remote locations away from the process equipment.

The systems looked obsolete due to what are now considered primitive text-only user interfaces. More important, the eight expert systems covered the maintenance of older equipment in older plants. As this equipment was replaced, the expertise in the expert systems became obsolete.

The use of LISP in those systems imposed significant limitations in system maintenance. Without LISP programming skills, system users could not maintain the expert systems themselves to reflect evolving diagnostic requirements. Users were also reported to have perceived a lack of corporate support of these systems and thus chose to not pursue system maintenance by others.

The use of these original eight expert systems has diminished. This experience notwithstanding, knowledge-based technology is reported to play a continuing role. A system called SIMON has been developed under the Aion Development System. SIMON calls upon FORTRAN analysis routines and a product information database to help determine whether food involved in a cooker malfunction can be salvaged and still meet federal food handling regulations. A timely determination can have a significant cost impact by avoiding the unnecessary disposal of food. SIMON is reported to save days in data collection and decision making.

A new SON OF COOKER system is also reported to be underway. This system supports a flexible processing line designed to process multiple soup products at speeds that may be varied to suit requirements imposed further up the line. The new system uses sensor data directly rather than requiring operator entry of this information. Faster computation and perhaps the use of hypermedia (sound and static or moving images in addition to text and graphics) are also under consideration.

7.2.6 Power System Management

An expert system provides network management support for an electric utility.

The Consolidated Edison Company of New York (Con Edison) electric power network covers 604 square miles in and around New York City. Con Edison handles loads of over 10,000 megawatts and interconnects customers with a variety of Con Edison generators as well as neighboring power grids.

The electric utility system is monitored by a real-time energy management system called SOCCS running on four Gould/SEL 3287 mainframe computers. This system provides a variety of network management functions, including load management, transmission line security analysis, and automatic generation control. Input to SOCCS is provided by 69 remote telemetry units (RTUs), which collectively provide SOCCS 100,000 data points every 2 seconds.

An expert system called SAA (SOCCS Alarm Advisor) was developed to support maintenance of the Con Edison system by monitoring events and alarms generated by SOCCS. Events include messages from operators and other utilities, manual data entries, and terminations of alarm conditions. Alarms are generated by tripped circuit breakers, telemetry failures from RTUs, and deviation of network electrical parameters from preset limits.

SAA performs two roles. The first is to separate alarms requiring attention from nuisance alarms. A typical nuisance alarm is caused by a poor connection on the telephone line connecting SOCCS to one of the RTUs. This causes the over 200 data points from this single RTU to rapidly toggle between normal and alarm states. SAA considers an alarm to be a nuisance if it changes status more than four times in six seconds. When such a condition is detected, SAA displays the alarm, logs it, and then removes it. This removes clutter from the display and makes legitimate alarms more visible.

SAA's second role is to analyze legitimate alarms to determine the underlying equipment failure. SAA contains logic to resolve data and alarm discrepancies. Repair recommendations are provided to return the system to a stable state. SAA recognizes and responds to circuit breaker failure, and provides identification of transformer spare bank and procedures for multiple feeder trip-out.

SAA design requirements included the ability to support real-time data rates without missing any alarms. This represents a challenge with an expert system approach and has been accomplished by prioritizing system actions. After filtering nuisance alarms, SAA displays legitimate alarms within 20 seconds, with more critical functions such as data input taking precedence over less critical functions such as responding to operator requests for information. Reported benefits include improved operator response time, standardization of operator responses to particular problems, a reduction of alarm clutter on operator displays, and prioritization of maintenance tasks according to problem severity.

SAA development began in September 1988. The system was in use since June 1989, with acceptance testing of the final version completed in November 1989. The vendor providing expert system development worked full time over the 15-calendar-month project, with approximately 5 man-months total provided by Con Edison domain experts and users.

This application illustrates strengths and limitations of expert system technology. Users receive specific advice conforming to Con Edison-approved procedures. However, a limitation is evident in speed. This is a problem in control and other manufacturing support applications. In this case, with features such as prioritization, the system is reported to handle alarms rapidly enough to meet real-time requirements. In the following application, an expert system was used that is specifically designed for rapid response. In both cases, the application was successful because, among other things, the application speed requirements were carefully matched to the expert system capabilities.

7.2.7 Petroleum Refining

Complex processing plants such as refineries present a challenging monitoring and diagnosis task. Out-of-tolerance operation, not to mention catastrophic equipment failure can impose significant risks. Timely and effective diagnosis is critical to cost and risk containment in such an environment.

Arturo Fermin reported the development of an expert system-based diagnostic system for an olefin alkylation process at Corpoven S. A.'s Puerto La Cruz refinery in Venezuela. This is a profitable but complex process involving hazardous materials—in particular hydrofluoric acid. A range of problems occur in a process of this complexity. Relatively minor process deviations are common and require constant operator attention. Even though most of the processes exhibit slow reactions, the interactions between processes are complex. Thus, troubleshooting and correction are often complex and time-consuming.

An operator with 20 years of experience can diagnose a typical problem in one hour. The process typically responds to the corrective action in two hours. Thus, on average, each minor problem generates a total of three hours of unacceptable product. Less common but more serious problems also occur. These may require that product be reprocessed or even that the line be shut down and restarted, resulting in several days of lost production.

The problem-solving challenges and the resulting costs created interest in improved troubleshooting capabilities. The imminent retirement of two operators with 38 and 40 years of experience provided an impetus to capture this valuable troubleshooting knowledge. The decision was made to develop an expert system-based diagnostic system. The system was to provide more complete diagnosis, calling upon the captured expertise. This expertise was also to provide more rapid diagnosis.

The problem-solving speed was considered important to the success of the diagnostic system. Shortening diagnosis by one hour would resolve some problems before they grew to the serious stage that was occasionally encountered. In addition, a one-hour reduction in problem-solving time was estimated to significantly improve the to-be-developed system, raising the projected effectiveness from 35 percent to 75 percent of problems to be resolved before product properties exceeded acceptable limits.

This requirement for speed represented a challenge, since expert systems are typically slow in response. For this reason, one element of the system specification was that, as much as possible, it was to be optimized for speed. The system was to operate on-line and thus have immediate access to data, rather than in batch mode with resulting delays in data access and response. The G2 Real Time Expert System by Gensym Corporation was chosen as the development platform for this application. G2 is inherently designed to provide the desired rapid response. In addition, bridge, or software connections are offered to link G2 to common PLCs (programmable logic controllers) and DCSs (distributed control systems), as used in this application, and thus provide real-time data access.

To further enhance the system's ability to provide timely advice, it was given forecasting capabilities. The knowledge base contained rules that, when provided data from the existing DCS system, enabled the troubleshooting system to infer the current process state and the trajectory, or direction and rate of change, of the process state.

Knowledge acquisition was performed by system developers working with the in-house experts, who provided problem identification methods, problem-solving logic, and patterns of plant operation and failure. Differences of problem-solving opinion were identified and resolved, sometimes by testing various approaches. The diagnostic method involved monitoring of visual cues as well as key operating parameters. The resulting knowledge base was designed so that an out-of-spec condition on one or more key parameters would trigger a sequence in which chains of rules would be involved to prove or disprove whether a particular root cause could generate the observed condition. Rules would also alert the operator to other problems that could be anticipated on the basis of the root cause.

The troubleshooting system was reported to forecast most process deviations approximately two hours before product quality limits were exceeded. Problems arising from human error or catastrophic failures such as ruptured pipes could not be predicted. However, better control of process deviations was reported to minimize the incidence of error and equipment failure by keeping equipment more consistently within intended operating ranges and therefore under less problem-inducing stress.

Presentation of problem-solving information must be carefully planned. The information must be accurate and as complete as possible. However, it must also be appropriate to the needs of the user. Different users will have different skills and interest levels and will confront problems ranging from obvious to obscure. The system must inform but not overwhelm.

For this reason, the system's graphical user interface was arranged in a hierarchical manner. At the top level, the alkylation process was represented schematically as boxes representing unit processes interconnected with arrows. Out-of-spec unit processes were indicated by a red dot in the appropriate box. More expert users may diagnose and resolve problems based on this input alone. Less expert users or more subtle or complex problems may require supporting detail available on lower-level screens. Screens providing the next level of detail incorporate cause-and-effect diagrams, logically linking observed problems to root causes, with problems and causes represented as color-coded circles. A user may then select one of the circles to observe detailed written instructions for that particular problem.

The following benefits were reported:

- The development of the knowledge base forced consensus to be reached in problem identification and resolution. This provided a useful body of knowledge, captured in the expert system and supporting charts, diagrams, and blueprints. This knowledge was expected to support tasks beyond plant operation, such as training, industrial security, and facilities upgrades.

- The expert system incorporated valuable expertise of veteran operators. This expertise was made available 24 hours per day, making the plant less dependent on the skills and availability of the operators on duty.

- The expert system provided a valuable training tool. New operators learned more than proper ranges for each measure parameter. They could view the process in a larger sense, seeing the interconnections between the various unit processes making up the larger process. This has improved their diagnostic and thus problem-solving skills.

7.3 NEURAL NET APPLICATIONS

7.3.1 Power System Security Assessment

Neural nets have demonstrated benefits in electric power system management. The ability of neural nets to provide a virtually instantaneous response is particularly useful in time-critical applications such as this.

In power systems, a particularly time-sensitive application is that of security assessment. When a failure occurs in a conductor or a piece of equipment in a network, the failure may cause the system to go into an unstable state unless corrective action is taken. This unstable condition, such as an increase in generator speed, can have catastrophic consequences. The security assessment task is to determine, for a given equipment or network failure, whether the system will remain stable.

In some cases, this assessment must be done very rapidly so that corrective action can be taken before the system fails. Traditional security assessment methods call upon massive computer systems simulating various network failure modes and calculating the effects on system security. This computer-intensive approach does not always provide timely advice. For this reason, research topics in power systems include faster security assessment methods.

Neural networks have shown promise in this area. Dr. Yoh-Han Pao and Dr. Dejan Sobajic at Case Western Reserve University (Cleveland, Ohio) have demonstrated a neural net-based security assessment to be at least 1,000 times faster than conventional approaches. The demonstration was performed on a simulation of a power system under development by NASA Lewis Research Center for the NASA Space Station. The prototype power grid contains two generators (current sources) and four loads (current sinks). The security assessment of this system was accomplished in 1.88 milliseconds with a Functional Link Net approach, and 2.0 seconds with the traditional load flow analysis, both running on a PC/AT personal computer.

Although solution times can be expected to scale up for larger networks, faster computers may be employed to limit the impact of larger networks with more failure modes. The pattern recognition nature of neural nets is the basis for the demonstrated increase in speed. The use of low-cost computers in this proof-of-concept demonstration points the way to practical neural net-based solutions of time-critical diagnostic tasks.

7.3.2 Continuous Casting

A neural net-based process monitoring system detects continuous steel casting process faults. It is able to detect the first stages of this fault so that compensating actions may be taken and in this way correct the situation and avoid quality and maintenance problems. This application demonstrates neural net-based process moni-toring capabilities, while providing maintenance and quality benefits for this new metalworking process.

In traditional steel making, molten steel is poured into molds, and the steel solidifies into ingots. These ingots are subsequently transformed into usable shapes such as plates or sheets by reheating and then rolling the ingots into successively thinner sections.

Continuous casting is a new steel-making process that avoids the wasted energy and inventory holding costs that result from the traditional practice of allowing a molten ingot to solidify, only to reheat it for further processing sometime later. Instead, the molten steel is solidified in a continuous process and only to the degree that the solidified skin over the still-molten core allows the material to be handled. Both solidification and subsequent rolling are per-formed continuously rather than on a discrete ingot-by-ingot basis.

Continuous casting is a sophisticated process. The molten steel enters a water-cooled mold. The outside surface of the continuous slab of steel gradually solidifies but must be kept continuously moving and contained within the walls of the continuous caster.

On occasion, a processing defect called breakout occurs. Break-out is a tear in the solidified skin of the steel as it moves through the mold. If this tear propagates too far across the solidified surface, molten metal will leak out into the continuous casting equipment. This requires expensive shutdown of the line.

A neural net-based process monitoring application was devel-oped and installed in 1990 by Fujitsu and Nippon Steel in Japan. After fine-tuning, the system is reported to detect all cases of break-out, to provide fewer false alarms, and to reduce costs by millions of dollars per year.

Before breakout has progressed to the point of requiring shut-down, there is evidence of its presence in the form of a localized hot spot on the surface of the mold. The hot spot is caused by metal from the hotter interior breaking through the cooler skin of the stream of solidifying metal. This hot spot is detected by thermocouples that measure mold surface temperatures and neural nets that determine whether the pattern of hot spots matches that of a developing breakout. In particular, neural networks analyze the thermocouple outputs to determine if a hot spot is propagating across the surface of the solidifying steel.

Thermocouples are arranged in two rows perpendicular to the flow of steel. Each of the thermocouples in the first row is connected to its own neural network. Ten consecutive temperature readings are taken for each thermocouple. These data points are time-shifted and stored so that all the readings from all of the first-row thermocouples are presented to their respective neural nets at once. The nets review this time-series data, looking for a hot spot moving across the thermocouples. The system looks for temperature gradients, not any particular temperature. The potential presence of the start of a breakout is thus established.

The next set of neural nets determines the location of the potential tear. Six successive readings are taken from each first-row thermocouple. These data are also time-shifted and stored so that the largest of the six readings from each thermocouple is available. These largest readings from each pair of adjacent first-row thermocouples are fed to two net inputs. The output of these nets indicates the location of a possible breakout.

Whether this will indeed become a breakout requires determination of the propagation of this hot spot. This is done by a similar arrangement of neural networks connected to the second row of thermocouples. Comparison of the location and size of the hot spot as it passes the first and then the second row of thermocouples enables the system to judge whether a tear is propagating in a way that will cause a breakout.

Development of the system required installation of the thermocouples and development and training of the neural nets. The nets in this system were reported to require 34 events for training, including 9 breakouts. The system was tested, with a perfect score, on data representing 27 events with 2 breakouts. The system was retrained during initial usage but was judged very successful, with plans reported for installation in other steel plants.

Common threads are evident in all of the applications in this chapter. Expert systems have demonstrated benefits in many monitoring and diagnostic applications. They provide structured logic and explicit reasoning, both useful in these applications. Neural networks have also demonstrated benefits in monitoring and diagnostic applications. The pattern-recognition approach provides rapid calculation, as demonstrated in power system security assessment. The casting application illustrated the ability of neural nets to learn how to detect abnormal situations. In most applications, these knowledge-based technologies are combined with traditional approaches to provide complementary analytical capabilities, and to integrate these capabilities into larger systems and sources of data.

This chapter has covered applications in which the goal is to verify that things are proceeding according to plan. In the next chapter, the focus shifts to determining the best process. The goal of process optimization—fine-tuning a process to best achieve objectives—is reviewed, and knowledge-based applications are presented.

7.4 REFERENCES

1. Barnes, Patrick O., and Hollis A. Hohensee. "An Expert System for Troubleshooting Resistance Welding Controllers." Proceedings of the 3rd Annual Expert Systems Conference, Engineering Society of Detroit, 1989, pp. 51–58.

2. Bartos, Frank J. "AI Now More Realistically at Work in Industry." *Control Engineering,* July 1989, pp. 90–93.

3. Bennet, Randy, and Alan K. Carr. "COOKER: What's Happened Since?" *Handbook of Expert Systems in Manufacturing.* Edited by Rex Maus and Jessica Keyes. New York: McGraw-Hill, Inc., 1991, pp. 330–37.

4. Gensym G2 product literature, Cambridge, Massachusetts: Gensym Corporation.

5. Hammerstrom, Dan. "Neural Networks at Work." *IEEE Spectrum,* June 1993, pp. 26–32.

6. Keyes, Jessica. "Manufacturing Survey." *Handbook of Expert Systems in Manufacturing.* Edited by Rex Maus and Jessica Keyes. New York: McGraw-Hill, Inc., 1991, pp. 42–55.

7. Parker, Kevin. "Lending Expertise to Supervisory Control." *Manufacturing Systems,* August 1993, pp. 24–28.

8. Pinto, James J. "Process Industries-Trends in the Nineties." *Automation,* September 1990, pp. 40–41.

9. Rasmus, Dan. "AI in the '90s: Its Impact on Manufacturing." *Manufacturing Systems,* January 1991, pp. 32–39.

10. Rowan, Duncan A. "On-Line Expert Systems in Process Industries." *AI Expert,* August 1989, pp. 30–38.

11. Silverman, Steven, and Alvin Shoop. "A Real-Time Expert System in the Area of Energy Management." *Handbook of Expert Systems in Manufacturing.* Edited by Rex Maus and Jessica Keyes. New York: McGraw-Hill, Inc., 1991, pp. 316–29.

12. Sobajic, Dejan J., and Yoh-Han Pao. "Neurocomputing in Power Systems." *INNS Above Threshold Newsletter.* March 1993, pp. 10–13.

13. Uhrig, Robert E. "Use of Neural Networks in the Operation of Nuclear Power Plants." University of Tennessee, Knoxville, Tennessee, and Oak Ridge National Laboratory, Oak Ridge, Tennessee.

14. VerDuin, William H. "Neural Networks for Diagnosis and Control." *Journal of Neural Network Computing,* Winter 1990, pp. 46–52.

15. Zygmont, Jeffrey. "AI Takes More Pointed Aim at Industry and Problem Solutions." *Managing Automation,* August 1993, pp. 34–35.

Chapter Eight

Knowledge-Based Process Optimization:
Combining Better Quality and Lower Costs

8.1 INTRODUCTION

Operating a process requires that, at a minimum, equipment be maintained in working order and process variables maintained within acceptable ranges. The previous chapter covered knowledge-based tools to accomplish these basic goals. In many processes, an additional knowledge-based opportunity exists. That opportunity is to find and implement ways to run the process better. The goal is to optimize the process to provide better quality, lower cost, or some other operational improvement.

This chapter addresses the increasingly important topic of process optimization. Competitive pressure requires manufacturers to pursue every opportunity to reduce costs and improve quality. As process optimization technologies improve, previously impractical solutions become justifiable. This chapter includes a review of the challenges in real-world optimization and the solutions provided by knowledged-based systems.

The goal of process optimization is similar to that of process monitoring and diagnosis, in that both seek to improve the effectiveness of manufacturing operations. But the nature and objectives of the two are quite different.

Processes are monitored to identify deviations from anticipated conditions. Monitoring is performed by the machine operator or maintenance staff with increasing help by computer-based tools. The task is to keep it running and to keep all parameters within predefined ranges. Monitored parameters may include both controlled process variables and measures of process capability, such as quality and production rate. If these measured parameters are found to exceed predefined bounds, diagnosis is performed to identify underlying causes, at least to the degree that is required to identify an appropriate corrective action. The best situation, from a keep-it-simple perspective, is one in which the process is relatively insensitive to variations in controlled and uncontrolled parameters. If this is indeed the case, there may be no point in considering additional complications such as the effects of equipment age and maintenance history, the impact of uncontrolled parameters, and perhaps the presence of interactions between parameters.

But many real-life situations involve those complications, and they have a measurable effect on the performance of a process. The premise of optimization is that, in these cases, an understanding of these effects and interactions can be used to better meet process objectives. This more sophisticated strategy may involve more sophisticated control hardware. But more commonly, it involves better use of existing hardware based on better understanding of the process.

Traditionally, developing this understanding required a time-consuming and perhaps disruptive series of tests. Tests would be done under a variety of conditions and would explore the effects of changes in setpoints, individually and in relation to each other. These tests might reveal whether better control strategies would yield better results. The temptation, of course, is to assume that they would not, even if this assumption is unwarranted, just to avoid the need for time-consuming experimentation.

Discovery of a better control strategy may be done by trial and error or by a more rigorous engineering study. The trial-and-error method may provide optimization insights quickly, particularly if trials are guided by an understanding of process interactions. Or, the trial-and-error may drag on with no way to know when, and if, the insights that will lead to better performance will be achieved. The time requirements for a more rigorous study based on traditional methods will be known in advance and will be large.

Often, the major argument against such expenditure of time is that the investment cannot be justified because the payoff may be too small or simply unknown.

The argument in favor of process optimization is that most processes can be improved, if only incrementally, and that somewhere a competitor will make that improvement if it hasn't done so already. That competitor will then enjoy a quality or cost advantage. Because of this, optimized processes are becoming a requirement to maintain, not just increase, market share.

To meet the needs of increasing competition, optimization must become more practical, more accessible. One way to do this is to streamline the analytical process used to determine optimization payback and approach. If this can be done, it will be possible to estimate payback early in the project, and on this basis make an informed decision on whether to spend the time to proceed with the analysis. The payback analysis will also be more favorable, because the time required to develop the optimization strategy will be lower.

Developing an optimization strategy is often time-consuming because the number of interacting parameters is so high. Process objectives may involve a number of dimensions. There may be trade-offs in product function and quality to be considered. Cost elements may include raw material, energy, and by-product disposition costs. Equipment durability and maintenance costs may be involved. The degree to which these perhaps competing objectives are met may depend on the interaction among controlled variables. It may also depend on uncontrolled variables such as ambient temperature and humidity. And the relative importance of these objectives may vary from day to day or product to product.

Traditional process optimization approaches based on first principles and statistical methods are effective, but at a cost that limits application to only those processes most critical to control or for which paybacks are most compelling. The good news is that knowledge-based technologies have demonstrated process optimization capabilities that make optimization faster, easier, and justifiable for more processes.

The following sections outline both traditional and knowledge-based process optimization techniques and applications. Common to all approaches is the need to discover how process conditions relate to process results.

8.2 THE NEED: A PROCESS MODEL

The objective of manufacturing is to best meet multiple customer needs at minimum total cost. A manufacturer's success in meeting often conflicting requirements plays a key role in determining long-term success. Each objective is important, yet many of them are in conflict.

There is no single best approach. Each manufacturer faces slightly different challenges and opportunities. Each application presents a different ranking of multiple objectives and a different degree of difficulty in meeting each.

Laws of chemistry and physics and the characteristics of specific products, processes, and operating environments define process results and the degree to which control objectives can be met. To best balance conflicting objectives and make best use of available opportunities, a process should be operated in a way that takes maximum advantage of interactions among parameters. To do this, the process engineer needs to know how the various parameters interact.

This understanding can come from experience or trial-and-error experimentation. But the best results can be anticipated if this understanding is explicit, in the form of an accurate, complete process model.

In some cases, this model can be derived from first principles. A first-principles model describes the cause underlying changes in process results. As previously described, in many real-world situations, this approach is not sufficient. And so, a model must be based on observed effects. These models capture relationships evident in data in which an observed effect is presumed to be the result of an assumed observed cause.

The traditional approach is statistical methods, which have a long and successful track record. However, opportunities exist to improve upon statistics-based capabilities and ease of use. Knowledge-based systems, in particular neural nets, have demonstrated compelling process modeling capabilities.

With each of the approaches, the process model is a mathematical relationship between process inputs such as setpoints, uncontrolled variable values, and raw material characteristics, and process outputs such as product characteristics and quality and the various elements of cost. Since this model is the basis for process optimization, its accuracy and completeness directly affect optimization results.

The choice between modeling methods thus involves accuracy as well as ease of development and maintenance. These factors are affected by the user's experience as well as innate attributes of the technologies. The following sections describe traditional statistics-based process optimization approaches. Knowledge-based approaches are then presented.

8.3 STATISTICAL APPROACHES

8.3.1 SPC and SQC

The connection between statistics and process monitoring is clear, with statistical methods developed specifically for this application area. Techniques such as **statistical process control (SPC)** and **statistical quality control (SQC)** are becoming commonplace. Progressive manufacturers are training plant floor staff to collect and analyze process data, calculating the means and distributions of the measured values of individual process or quality parameters. Plotting these values over time provides an indication of trends. From this, operators can predict out-of-bounds conditions and initiate correction of the problem.

Process optimization represents the conceptual next step, moving from identification of changes to developing a detailed understanding of relationships between process inputs and outputs. This understanding is captured in process models, and it enables process engineers to find the best ways to meet specific, and perhaps multiple, process objectives. Statistical methods support process optimization as well. The Taguchi method is a common approach.

8.3.2 The Taguchi Method

This process optimization method was developed and applied by Dr. Genichi Taguchi in the 1940s. An early success was to help rebuild the Japanese telephone network after the war. The problem-solving challenge was to reduce the task of troubleshooting a massive and essentially inoperable network to manageable size, so that restoration could proceed as rapidly as possible. Taguchi's approach involved simplifying assumptions that reduced the number of experiments required. The rapid restoration of the network demonstrated the merits of this method.

The Taguchi approach is now most commonly used to determine the least expensive way to achieve required levels of product functionality and quality. As a quality improvement method, it seeks to maximize quality while minimizing the cost to provide that enhanced quality. The Taguchi method can therefore be viewed as a process optimization technique.

The stated goal is to "minimize losses to society" that occur when product quality is less than the best achievable. Dr. Taguchi broadly defines these losses to include not only in-process repair and warranty costs, but also loss of reputation due to customer dissatisfaction.

The Taguchi method is presented as an alternative to improving quality by the traditional approach of buying better raw materials, processes, or controls. The Taguchi alternative is a sensitivity analysis in which the effect of parameter variations on product quality is determined experimentally. The point is to cost-effectively improve quality by implementing the least expensive quality-improving changes first.

Once the expense of the Taguchi experimentation and model building has been absorbed, the model may reveal low-cost or even no-cost quality improvement opportunities. No-cost improvements may be possible if the model reveals that quality will be improved by changes in the target values of certain process parameters. In other cases, the model may reveal that process parameters must be more tightly controlled or that raw material properties must be improved. Both of these actions will incur costs, but the model-based knowledge of the relative impact of each parameter will enable a given quality improvement to be made by improving the parameter that yields the greatest improvement per dollar spent.

Determination of the most effective ways to vary controllable parameters includes an assessment of the effects of those parameters that cannot be controlled and, in Taguchi parlance, are considered **noise.** Noise includes so-called *outer noise,* that is, noise outside of a part, including environmental factors such as temperature, humidity, and vibration. *Inner noise* consists of changes that occur within a part due to, for example, wear or rust.

The Taguchi method is widely used because it has been demonstrated to work. It is particularly useful as a screening technique, making the rough cut to find the most important factors causing a severe quality problem. In this type of application, the experimental compromises made in the interest of saving time are often of minor consequence.

The Taguchi method involves simplifying assumptions made to minimize experiments and thus provide faster results. If a given parameter is affected by seven factors, classical statistics would require that $2^7 = 128$ experiments be run to fully characterize the relationships between the seven factors and the output parameter.

The Taguchi method is a fractional factorial method, meaning that a fraction of the many possible factors, or interactions, are investigated. In the above case of a seven factor problem, $2^{7-1} = 2^6 = 64$ experiments would be required.

The savings in experimentation means that interactions between parameters are ignored, that is, assumed to be of negligible consequence. This assumption may or may not be valid, and the Taguchi method cannot determine which. A final validating experiment is performed to verify that observed variations in the output can be fully explained by variations in the individual inputs. If sufficient correlation is not found, the Taguchi method will conclude that the process is not controllable and should be redesigned.

That conclusion may be valid, but there are other possibilities. All factors may not be independent, and so it may not be valid to ignore interactions. Most processes involve factors that interact at some level. Ignoring these interactions may be valid as a first cut approximation. But in cases where the big factors limiting process capabilities have already been resolved and the last few percentage points of quality or efficiency improvement are to be found, these interactions can be critical.

In these cases, traditional, more rigorous statistical approaches such as regression analysis are preferred because they are better able to uncover such relationships. Of course, these methods require a greater number of experiments or data points. In modeling, there is no free lunch. The model can represent only those relationships made evident in data. This means that data must be collected in a way that reveals all interactions between all variables.

When considering how much data must be generated, or whether available data are sufficient, the above sentence may not appear particularly helpful. Unfortunately, this is the only absolute statement that can be made about data requirements.

Moving from theory to practice, questions arise regarding easy ways to generate data and ways to minimize the need for data. In principle, data can be collected from the actual process to be modeled. This has the advantage of providing plenty of data with virtually no effort required to generate them.

There are disadvantages with this approach, particularly in support of statistical methods. One problem is that the data tend to be masked to some degree by noise, random variations that have to do with the data collection, not with the process itself. With statistical methods, this noise must be explicitly removed and still may introduce error into the model.

Another problem with operational data is that if the system is reasonably well controlled, data exhibit relatively narrow ranges of variation. More interesting data are available only if steps are taken to run the process and generate data over wider ranges. Without these explicit steps, the model's validity is limited to narrow ranges, raising the possibility that better control strategies outside of current limits will not be discovered.

The model-builder's enthusiasm for special data-generating trials on production equipment is rarely shared by plant management. Most plant managers are reluctant to allow time-consuming experimentation on production equipment, even without the risk of surprises due to extreme parameter settings. For these reasons, the preferred (if not always feasible) approach is to base process models on experimental data. These data can be collected under controlled, noise-free conditions. Data points can be selected to represent extreme values of individual parameters, so that the resulting model captures all interactions over a suitably wide range.

8.3.3 Design of Experiments

Given a willingness to conduct experiments, the question remains of how to do so most effectively, providing the most useful data with the fewest experiments. A technique called **design of experiments (DOE)** is used to suggest appropriate experiments for a given application. The user identifies the ranges of each of perhaps many parameters. The DOE technique then provides the user a list of experiments to be run that will test the response of a system under all combinations of extreme parameter values (and for intermediate values as the testing budget allows).

This technique is supported by stand-alone DOE software packages, as well as DOE modules in statistical software packages available from vendors such as SAS and BBN. Common DOE capabilities cover a broad scope, from screening experiments that determine the most important factors in a given situation, to experiments that explore two-factor interactions. These two-factor interactions can be displayed graphically as response surfaces in which the horizontal axes represent input factors of interest and the vertical axis the response, or output parameter of interest.

Designing experiments to fully explore relationships is a valuable task, whether the test results are the desired end results or whether they are used to support development of a model. This is true whether the model is based on statistics or neural nets. A potential source of confusion is that the term DOE is sometimes used to refer to both experimental design and statistical model development as one task. In the balance of this section, the term DOE is used in this way, to signify a task combining design of experiments and statistical analysis of these results.

Under this definition, DOE is subject to the limitations of statistics, with limitations based on data quality presumably overcome by use of proper experimental design. As a practical matter, however, DOE may ask for more than can be delivered. Many testing budgets or timetables do not allow a rigorous DOE-based testing program.

In other cases, it can be difficult or impossible to provide the DOE-recommended samples. The adhesive formulation example presented earlier involved interdependent process variables. This allowed the variables to change only in certain combinations and precluded the production of products with, for example, one parameter at the low limit and the other at the high limit. The use of statistics was constrained by this inability to independently control these parameters, and so in this example neural nets were the preferred approach.

Another DOE limitation, also described in the chapter on formulated product design, is that the technique is relatively intolerant of missing data. Neural net-based approaches were shown to be more tolerant of gaps in data, as well as more accommodating of mixed types of data and data from experiments that did not reach the desired conclusion. Similar issues arise in discrete part manufacturing. Providing the samples required by statistical methods may involve extreme measures. Each sample part may require special tooling, with each combination of parameter values (dimension X at its high limit, dimension Y at its low limit, dimension Z at its midpoint, and so on) requiring a unique set of tooling. Thus, many sets of potentially expensive special tools and fixtures may be required.

These model development cost issues, based on the time and cost of experimentation, provide the incentive to develop new modeling methods. The following section on knowledge-based approaches illustrates the ways that knowledge-based technologies extend capabilities in process optimization.

8.4 KNOWLEDGE-BASED APPROACHES

Several knowledge-based technologies have been applied to process optimization. Expert systems, neural net, and integrated neural net/optimizer approaches have been reported. In fact, computer-based process monitoring and optimization incorporated expertise long before terms such as expert systems or AI existed. Any control strategy that incorporates problem-solving expertise can be considered knowledge-based, whether that expertise is in the form of expert system rules or Fortran IF-THEN statements.

What's new is the range of techniques available. The most widely implemented knowledge-based technology, expert systems, has been implemented as a process optimization approach. Process experts identify rules that optimize a process under given circumstances. The expert system acts upon these rules, resetting process parameters to suit current or anticipated conditions.

The limitations of this approach are those of expert systems in general. Speed may be an issue, particularly for systems with rapid state changes. This limitation is not relevant in slow-changing systems. In faster-changing systems, this limitation can be overcome by configuring the process optimizer as a stand-alone advisor that supports but is independent of the real-time controller.

The larger issue is that of understanding how to optimize the process. The complexity of many processes, including multidimensional interactions, nonlinearity, and time-varying behavior, makes the development of effective and complete rules difficult if not impossible within available budgets and timelines. In process optimization as in other areas, a need exists to discover the relationships that will support a process optimization model.

8.5 NEURAL NET APPLICATION: ORGANIC CHEMICAL PRODUCTION

Neural nets have demonstrated capabilities in this area, based on their ability to discover relationships in data and adaptively improve their performance. These capabilities are useful in applications such as process optimization that are characterized by complex and evolving system characteristics. The following application illustrates typical benefits. In particular a neural net-based process model was developed to predict product properties and the accuracy of this model compared to a theoretical (first principles) model and to experimental data. The neural net model was found to be at least as accurate as the theoretical model.

John Blaesi and Bruce Jensen of Johnson Yokogawa, a major controls supplier, have published a comparison of neural net and first principles modeling, based on applications in support of organic chemical production. The objective was to predict properties for two products produced by a coking furnace combination tower. The properties of the two products are characteristic temperatures—in particular, coker naphtha end point and coker furnace oil 90 percent point.

The rationale for developing a neural net model was to save time in system development and validation and to minimize the need for an expert in coke fractionation theory to support system development and maintenance. The self-learning and adaptive capabilities of neural nets were to provide these savings, although neural net validation by an expert would still be required. It was also anticipated that the computationally faster neural net-based model could provide more timely advice based on more current data than a first principles model.

Potential problems that were anticipated with the neural net approach related to data requirements, in particular the need for large quantities of data and for data covering a wide range of operating conditions. In this case, the presence of a plantwide information system and the availability of several years of operating data were helpful. Acquiring the hundreds or thousands of records required to train and validate the neural net might otherwise have been a problem, since new records would be generated at the rate of several per week.

In the course of developing the neural net model, it was discovered that of the 29 parameters thought to be important and therefore incorporated in the first principles coker fractionator model, only 14 were dominant. The neural net model based on these few parameters was found to at least equal the accuracy of the first principles model.

The inferred product properties, as established by the first principles and neural net models were compared to actual values obtained by laboratory testing over a 50-day period. It was noted that laboratory test conditions introduced considerable variability or noise into readings with 10° F errors encountered in measurements ranging from 330° to 580° F. This meant that an absolute benchmark did not exist for comparison of the two modeling methods. However, the following conclusions were reached:

- Both the first principles and the neural net model predicted changes in process direction and did so essentially instantly. This contrasted with the 12 hours otherwise required to accomplish lab testing and enabled far more timely correction of process deviations.
- Both models exhibited less variance than measured properties and thus appeared to be successfully filtering the noise present in measured values.
- The neural net model was believed to be at least as accurate as the first principles model and was faster and easier to develop.

These results validate neural nets as a modeling approach with benefits over alternative approaches. This is significant in the context of process optimization, due to the critical role played by the process model.

The remaining element to be added is to link the neural net-based modeling capability to an optimization capability. The challenges in accomplishing this were described previously in the formulated product design application. These challenges were also addressed in the following process optimization applications, also using commercial *integrated neural net and optimizer* tools.

8.6 INTEGRATED TECHNOLOGY APPLICATIONS AND PAYBACK

8.6.1 Petrochemical Production Optimization

The Eastman Chemical Company is a major manufacturer of chemicals, fibers, and plastics. Their Texas Eastman plant in Longview, Texas, uses over 30 individual chemical plants, staffed by over 2,800 employees, to make a variety of petroleum-based products. As reported by Ken Denmark, Mike Farren, and Bill Hammack of Texas Eastman, these plants provided opportunities for improvement common to many process plants. An integrated neural net and optimizer software product called Process Insights from Pavilion Technologies, Inc. was used to achieve clear-cut operational benefits.

Although large quantities of process data may be available, questions may remain unanswered. These questions may be very specific: "Can I believe this sensor reading?" Tactical questions might include: "Is there a better way to operate this plant?" The answers to these questions can have a significant impact on quality and costs. Yet, even with the assistance of voluminous data and skilled staff, these answers often cannot be fully answered. The challenge is to develop the detailed understanding of a process that will answer these questions in a cost-effective and maintainable manner. This understanding requires more than knowledge of acceptable setpoints. It requires an understanding of the interactions between the various controlled and uncontrolled process parameters.

This depth of understanding is not absolutely necessary. A basic understanding of the process, along with a fair amount of trial and error, will enable many processes to be tuned within, say, a few percent or even a few tenths of a percent of their optimum. But a detailed and up-to-date understanding of process interactions is required to achieve the last few percent of process optimization. The benefits of this optimization may be measured in production rates, quality, or costs. Even for production quantities well under the scale of Texas Eastman, the benefits can be compelling.

An accurate and up-to-date process model can also provide benefits in areas such as sensor validation. Identifying sensor faults other than the most obvious requires some basis of comparison, some ability to judge what is an appropriate reading under the circumstances. In the absence of redundant sensors, a process model is required to infer an appropriate if approximate reading, based on other sensor readings and knowledge about the system state.

The discovery capabilities of neural networks have been demonstrated to be a useful model-building approach. The integration of optimization capabilities has been shown to be a powerful manufacturing support capability. Texas Eastman reported the following benefits from their applications:

- Sensitivity analysis provided insight on ways to retune processes. The recommendations were often surprising, and sometimes enabled improvements with little or no investment required.
- Setpoint optimization was implemented. Users specified process objectives such as production rates, yields, quality, and costs, and setpoints were identified that best met these specified objectives.
- The integrated neural net/optimizer was used primarily as a real-time advisor. The system monitored operations in real time and helped operators identify faults, predict process deviations, and retune the process.
- Using the system for **closed-loop control** was investigated. In this mode, the system would move beyond dispensing of advice into direct implementation of control. If this had been implemented, it would provide the benefit of reducing the need for operator involvement. However, it would impose far more rigorous system development and validation requirements, since by being less involved, the operator would be less able to identify and respond to process upsets and other unanticipated circumstances.

- *Virtual sensors* were implemented and provided estimates of critical but unmeasurable parameters. This reduced product testing requirements and minimized production of substandard product. Hostile environments, such as extremes of acidity or alkalinity, made direct sensing of some critical process parameters difficult or impossible. The traditional approach in these circumstances was to measure these or related parameters after the fact, through laboratory analysis of samples of the completed product. The sampling approach and the time lag imposed by this method would often result in the production of substandard product for several hours before a problem was identified and corrected. Texas Eastman used the process model, incorporating available sensor data, to estimate critical but unmeasurable parameters. In this way, virtual sensors identified quality problems and corrective actions.
- The process model was also used, in the same way, to validate actual sensors. The model estimated appropriate sensor values, based on other sensor values and knowledge of the system state. By providing a reality check on sensor readings, the model detected less-obvious sensor failures.
- Preventive maintenance was enhanced by an improved ability to predict periodic maintenance needs. Some components require periodic maintenance to maintain effectiveness. As an example, heat exchangers in some applications require cleaning to prevent fouling, or blockage. In this application, Texas Eastman taught a neural net the changes in overall process characteristics due to gradual fouling of the heat exchanger over the course of its maintenance interval. This learned relationship enabled the process model to estimate from real-time data the progression of heat exchanger fouling and therefore to determine when service should be scheduled.
- In one chemical reactor optimization application, it was discovered that a less-expensive blend of chemicals could be used as input to the process without degrading the quality or quantity of the resulting product. This result was replicated on similar reactors and provided a considerable cost savings without the need for capital expenditures.
- In another optimization application, the neural net-based process model enabled further optimization of a process already the target of considerable optimization effort. The

objective was to reduce the use of an expensive additive that was required to chemically convert an undesirable process by-product to eliminate its effect on product properties. The use of the expensive additive varied over a wide range, since the by-product content in the product also varied over a wide range. In the course of 30 years of operation, the process parameters affecting by-product levels had never been discovered.

A neural net model discovered the critical relationship between process parameters and by-product content. The model incorporated 30 process variables and found that two variables were key. A sensitivity analysis revealed that adjustments of setpoints for these two variables, a no-cost change, could reduce additive usage by about 30 percent. The savings were realized after model validation procedures verified the predictions.

8.6.2 Investment Castings Yield Improvement

A manufacturer of investment castings identified an opportunity to implement process modeling and optimization, with the goal of decreasing the scrap rates of their investment cast products. These products are subjected to extreme thermal and dynamic loads. They are expensive to make and, therefore, expensive to scrap if they do not meet stringent requirements.

High-temperature alloys are used and near-net shape formed by investment casting into complex and precisely controlled geometries. To provide the required strength and durability, the crystalline structure of these parts is critical. Solidification of the high-temperature alloy must be precisely controlled to provide the required crystal size, shape, and orientation.

The incentive to improve yields is driven by the inability to retrieve much of the value added by the manufacturing processes. Beyond the costs of the specialized investment casting process, significant direct labor is required for machining of parts and inspection for conformance to dimensional and microstructural requirements. These casting, machining, and inspection costs are irretrievable when parts are scrapped.

The incentive to improve yields goes beyond cost issues. These parts are often produced in small lot sizes and may be subject to tight delivery deadlines. Low yields can present scheduling challenges by making less-effective use of production facilities for these short runs and by limiting the ability to maintain aggressive schedules.

Evidence existed that significant yield improvements were possible. The investment casting process is supported by state-of-the-art processing equipment and controls, but there remained an element of mystery regarding wide variations in process yields between various products and facilities.

The initial production of a new part design was often plagued by low process yields. Trial-and-error modifications of process parameters would eventually bring yields up to acceptable levels. But even for parts already in production, yields varied considerably. For a family of similar products produced in similar facilities, scrap rates varied by up to a 2 to 1 ratio.

The fact that yields could be raised to acceptable levels for certain products and plants after enough experience was gained suggested that the process was inherently controllable. The wide range in yields suggested that the process was not optimized and was, in fact, not completely understood. The goal was to gain that understanding and on this basis to improve yields.

It was determined that as a first step a process model would be developed, linking known process parameters to measures of product performance and quality. At a minimum, this model could determine whether variations in known process parameters or interactions between these known parameters could explain variations in scrap levels. The model might also help identify unknown but critical process parameters. This insight would drive a process optimizer that would identify better control strategies or other changes in manufacturing environment.

The CAD/Chem integrated neural net-based modeler and optimizer, as previously described in support of formulated product design, was selected for this application. In product design applications, this tool was used to model the effects of changes in product recipes on product properties and then to optimize product properties. In this application, a model was developed to show the effect of changes in processing parameters on product performance and quality. The optimizer then determined how to adjust process setpoints and procedures to optimize yield.

The inputs selected for the process model included a variety of foundry parameters that define the circumstances surrounding pouring of liquid metal into the array of investment casting molds. These parameters included

- Pour temperature—the temperature of the liquid metal.
- Preheat temperature—the initial temperature of the investment casting molds.
- Pour speed—a dimensionless variable that describes the rate at which the ladle of molten metal is rotated to pour the metal into the complex assembly of investment casting molds.

During initial process optimization work, a total of 31 input variables were used. The four outputs were measures of process yield, as indicated by the percentage of parts judged acceptable by the following four quality tests.

- *Grain size* is evaluated for conformance to limits after the grain boundaries are revealed by an alumina bead blast followed by acid etching.
- *Fluorescent penetrant inspection* and *X-ray inspection* are used to verify the soundness of castings, with shrinkage and porosity not to exceed acceptable limits.
- *Dimensional inspection* is performed with mechanical devices and electronic coordinate measuring machines to verify conformance to dimensional requirements.

The first round of process optimization was performed with a total of 215 records encompassing the 31 input and 4 output parameters. These records represented approximately six months of production data for the parts surveyed.

Data were entered into the modeler/optimizer by computer file transfer and manual entry as required. System setup called upon CAD/Chem data analysis, neural network training, and optimization screens. In this case, the optimizer objective was to maximize yield, and this was accomplished by maximizing yields of each of the four tests. Desirability functions reflected the-higher-the-better preferences.

Final results of this process optimization effort were not available, but preliminary conclusions included the following:

- Estimates from the neural net modeler within CAD/Chem did not match the available data. This indicated that there were unanticipated effects from process parameters that were not directly measured.

- This valuable insight initiated an investigation of the role of suspected critical but unmeasured parameters. Insulation wrapping procedures were found to be critical in controlling heat transfer and thus crystal growth and to vary from shift to shift. Once the problem was identified, the solution of operator training was cost free.

- Annual cost savings are conservatively estimated by the user to be $50,000, providing a payback period of under seven months based on the cost of the modeler/optimizer software.

The final payback numbers will be known only when the optimization study is complete. However, one might speculate that the lower-yield products and facilities could match the performance of the better products and facilities and that even if the performance of the better facilities remains unchanged, annual savings might be orders of magnitude higher than those estimated.

Conclusions that can be drawn from this application include the usual need for data. This application also illustrates the way the unknown and perhaps unanticipated relationships can be discovered. Paybacks can include quantifiable benefits in quality and other measures of manufacturing effectiveness, as well as improved insight on process improvement opportunities.

This application called upon a software tool designed for use by engineers. Sophisticated and flexible analytical and what-if capabilities, supported by 3-D response surface presentations, are designed to meet the needs and perspectives of well-educated and intellectually curious users. These users are commonplace in most engineering or scientific organizations. However, out on the plant floor, needs and perspectives may be more focused on immediate needs and required actions. These plant-floor perspectives are reflected in the following application.

8.6.3 Power Plant Combustion Optimization

Commercial electric power utilities supply power generated by a variety of methods. A major source of power is the combustion of fossil fuels such as coal, natural gas, and fuel oil. The combustion process generates steam, which in turn rotates a steam turbine to rotate a generator to generate electricity.

Although electric utility technologies continue to evolve, many 50-year-old plants are still in operation. These plants were built when fuel was inexpensive and readily available and emissions not a concern. At the time the primary design goals were reliability and low purchase and installation costs. These factors are still important, but they are now joined by the need to minimize fuel usage and control emissions.

Replacing these old plants is generally not a short-term option. Instead, the old plant must be given new capabilities. These capabilities may include the ability to use different, or even multiple, fuels to ensure availability of fuel and to meet emissions requirements. In addition, a plant designed to run at a constant load may now be called upon to support *swing* (time-varying) loads.

This change is caused by changes in the way electricity is transported and sold. Traditionally, electric utilities were essentially autonomous. They would generate and distribute all of the electricity used by their customers. These customers were located in a specific geographical area.

Now, electric utilities are interconnected with adjacent utilities. Power is bought and sold and transported across one or more utilities' networks to provide the least-expensive power when and where it is required. There is no longer any necessary relationship between the location where electricity is generated and where it is used.

Sophisticated power networks and transmission capabilities enable utilities to buy as much power as possible from the most efficient producers, which in general are the newer facilities. The older plants are still required to sustain peak loading, and so they remain operational. But swing-load usage represents a significant change from the usage for which the old plant was designed.

The operational characteristics of an old plant can also be expected to change with age. Wear and tear and equipment replacements cause changes in combustion and heat transfer processes. These performance changes are often not calculated or measured, and so control strategies do not reflect these evolving changes.

The need to accommodate changing load schedules while improving efficiency and minimizing emissions suggests that combustion processes be upgraded with new hardware. Although this is done in some cases and is certainly effective, it is often not an option. The question, then, is to what degree can combustion efficiency and emission levels be improved without expensive retrofits.

An opportunity exists to optimize combustion on the basis of an adaptive combustion model. This model would extract the maximum efficiency and minimum emissions possible without retrofits by basing control strategy on current circumstances. These circumstances include updated heat balances, equations describing the sources and destinations of heat in the combustion chamber and heat exchangers that incorporate current performance parameters, not those of years ago.

Extracting that last increment of performance also requires a real-time on-line system, that is, a system that immediately recognizes and responds to circumstances as they change. Some circumstances such as equipment characteristics change slowly. Examples of more rapidly varying circumstances are changes in the heat content of the fuel being burned and changes in the load on the generator, and thus the rate at which steam must be produced.

The need for an on-line adaptive model suggested the use of a neural net approach. The need to link that combustion model to an optimizer suggested the use of the technology platform underlying the CAD/Chem design tool previously described. It was felt that since this approach would provide clear benefits and paybacks to the electric power industry, a product opportunity existed. A product development and application partnership was initiated by AIWARE and Pegasus Technologies Corporation, a firm specializing in software development, installation, and support for control of fossil fuel-fired power generation systems. The following section describes the nature and application of this knowledge-based tool targeted to a narrow vertical market.

NeuSIGHT, or Neu(ral net-based process in)SIGHT, provides plant-floor staff recommendations on changes in control settings to improve combustion efficiency and thus reduce fuel costs, as well as reduce production of sulfur dioxide (SO_2), a contributor to acid rain. Another optimization objective is to minimize the production of oxides of nitrogen (NOx), another combustion by-product that causes smog. Each installation is custom configured to incorporate available sensor data and make best use of available hardware.

3 content:

Additional outputs may be added to the combustion process model, including but not limited to carbon monoxide (CO) and smoke opacity. These outputs are of interest as key measures of the combustion process. In addition, opacity is regulated to stay within defined limits.

Modeling and optimization capabilities are provided by a software system that incorporates essentially all of the functionalities of the technology platform underlying CAD/Chem, including the Functional Link Net-Based model generator and integrated optimizer. User interface screens similar to those of CAD/Chem are used for system setup and are used by process engineers for what-if studies to explore the effects of new control strategies or proposed hardware changes.

These screens also allow the engineer to model individual pieces of equipment. This is done to analyze a problem in general and each piece of equipment in particular. In this usage, the analysis is performed off line, in a nonreal-time mode.

The following features are provided to meet the special needs of this real-time plant floor application:

- Real-time data access.
- System manager to initiate data acquisition, neural net training, optimization, and display updates.
- Operator interface screens designed for easy plant-floor use.

Installation of the process optimizer does not require process shutdown. Installation tasks include providing access to sensor data, typically through a RS-232 link, system configuration and tuning, and customization of user interface screens. To realize the full optimization benefit may require the installation, in some cases, of additional sensors and related hardware and wiring.

A critical element for plant-floor systems is to keep them as simple to use as possible. This requires that data access not only be real time but also completely automatic, requiring no operator action once this stand-alone system is in place. The plant's existing data acquisition system is automatically polled.

The optimizer's system manager determines whether system characteristics remain sufficiently unchanged so that the existing combustion model is still valid and thus can continue to be consulted. If it is determined that the old model has become obsolete, retraining of the neural net model is initiated and completed within one minute. This determination is made by investigating clustering of current data compared to that of previous data. The optimizer then calls upon the new combustion process model to establish recommended changes in process setpoints.

Constraints, priorities, and desirability functions are set by a process engineer during system commissioning or as updates are required, using engineer interface screens similar to those previously shown for CAD/Chem. The constraints, priorities, and desirability functions may also be automatically adjusted in real time. This is done by predefined custom algorithms that adjust target values on the basis of input parameters such as load.

To keep things simple on the factory floor, operators never encounter the flexible, relatively complex engineer interface screens or any of the related capabilities of the full system. Instead, the operator interacts with the system solely through the simpler, more focused operator interface screens. These screens are designed to reduce information to the simplest meaningful level, providing the operator specific requests for action with supporting detail provided only as strictly required.

The primary operator interface screen is shown in Figure 8–1. The design intent of this screen is to tell an operator, who may be seated 12 feet away, what parameters should be varied and by how much to meet efficiency and emissions objectives. Let the engineers worry about multiple objectives, trade-offs, constraints, and so on. The operator probably doesn't care about these peripheral issues, but he or she wants to know what to do right now.

This screen provides the required input. The number of variables to be optimized varies from installation to installation, but this screen illustrates a six-parameter installation. In this case, the screen is dominated by a (green) six-sided figure. Each of the six edges represents one control parameter. In this case, main steam temperature, main steam pressure, excess oxygen, reheat temperature, feedwater temperature, and economizer outlet temperature are displayed. These parameters are to be controlled with the single objective of lowering heat rate.

The display shows at a glance that five of the six parameters should be adjusted to lower heat rate. The triangles emanating from each side of the figure indicate the magnitude and direction of

FIGURE 8–1
NeuSIGHT Operator Interface Screen

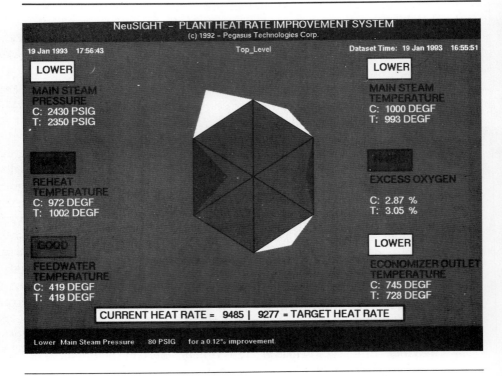

the recommended changes. Triangles pointing out from the figure indicate that parameters should be lowered, and vice versa. The height of the triangles indicate recommended magnitudes.

Text provides supporting detail. The current (C) and target (T) values are shown for each parameter. The efficiency improvements that are estimated for each parameter change are shown along the bottom. In this case, a 0.42 percent improvement is projected for the three parameters listed at the bottom of Figure 8–1.

Since this system runs on a dedicated workstation, operator interface screens can be displayed at all times. The operator may also choose to refer to other screens, such as a tabulation of parameters and associated upper and lower limits, or a bar chart presentation of the Figure 8–1 information.

Another screen helps operators manage process start-up, as well as the increasingly frequent need to shift from one output power level to another. There is a preferred relationship between main steam pressure and temperature that should be observed during power plant start-up, moving from ambient up to operating pressure and temperature. This same relationship holds when the temperature and pressure are to be moved from one level to another to vary power output.

This preferred relationship is indicated on a screen consisting of a graph of main steam temperature versus pressure. Current temperature and pressure readings are plotted to help the operator move the process along the displayed ideal trajectory of change in pressure per change in temperature. The operator can also verify that the process is not approaching the nonlinear boundaries of the overtemperature and overpressure regions that are also shown.

These capabilities help fossil fuel-fired power plant operators to better meet operating objectives. Management of start-up and changes in load are improved, as are steady-state efficiency and emissions. These benefits vary from installation to installation, but heat rate improvements of 0.5 to 5.0 percent have been demonstrated, as have reductions in NO_x emissions of 40 percent on coal-fired units without low-NO_x burners and of 25 percent on units with low-NO_x burners. The user can simultaneously optimize for both objectives, defining the trade-off between the two competing goals.

System installation and training costs vary but are generally in the range of $120,000 to $180,000 for a complete system. The heat rate improvement, and thus fuel savings alone, provides estimated paybacks of 6 to 12 months for typical medium-size (400 megawatt) plants. When factoring in direct reduction in SO_2 emissions, avoided capital costs for meeting NO_x requirements, and the possibility of selling SO_2 and NO_x emission credits, the expense is recaptured more rapidly.

This knowledge-based combustion optimization capability may also be applied to other process optimization applications. The fundamental requirements are that process data can be acquired and that the process can be optimized—that is, that some combination of changes in process setpoints can have a detectable impact on the degree to which process objectives are met.

As a practical matter, process optimization applications will probably be driven more by payback than technical feasibility. The expense of installing NeuSIGHT on fossil fuel-fired power plants is based primarily on the significant on-site services required to ensure that sensor data are available and automatically accessible. For the fuel and other savings possible, the expense is more than justified.

For other applications, installation may be simpler, costs lower, and payback shorter. Installation costs, particularly sensor data access costs will vary considerably from case to case. Since the decision to proceed with optimization is typically based on anticipated costs and benefits, these costs merit careful review.

The applications outlined in this and previous chapters demonstrate that knowledge-based systems have been successful in a wide variety of design and manufacturing applications. These applications point the way to many successful future applications in these and related areas.

The next chapter presents a view of trends in knowledge-based systems. This background on past successes and future opportunities sets the stage for the final how-to-do-it chapter, with the hope that the reader will be both encouraged and enabled to proceed.

8.7 REFERENCES

1. Babb, Michael. "Fast Computers Open the Way for Advanced Controls." *Control Engineering,* January 1991, pp. 45–51.
2. Bailey, S. J. "Using Alarms and Annunciators to Assess Fault Significance." *Control Engineering,* April 1990, pp. 101–3.
3. Bartos, Frank J. "Statistical Methods Team Up with Manufacturing Controls." *Control Engineering,* July 1990, pp. 63–66.
4. Brink, James R., and Sriram Mahalingam. "An Expert System for Quality Control in Manufacturing." *Handbook of Expert Systems in Manufacturing.* Edited by Rex Mous and Jessica Keyes. New York: McGraw-HIll, Inc., 1991, pp. 455–66.
5. Blaesi, John, and Bruce Jensen. "Can Neural Networks Compete with Process Calculations?" *INTECH,* December 1992, pp. 34–37.

6. Denmark, K., M. Farren, and B. Hammack. "Turning Production Data into SPC Gold." *INTECH,* December 1993, pp. 18–20.

7. Dumont, Guy A., Juan M. Martin-Sanchez, and Christos C. Zervos. "Comparison of an Auto-Tuned PID Regulator and an Adaptive Predictive Control System on an Industrial Bleach Plant." *Automation,* 25, no. 1 (1989), pp. 33–40.

8. Garg, Ashutosh. "Trimming NO from Furnaces." *Chemical Engineering,* November 1992, pp. 122–30.

9. Jones, Jeremy. "Identifying Models for Advanced Control." *Control and Instrumentation,* July 1992, pp. 30–31.

10. Kesley, Justin. "Application Watch: A Forge Scheduling System." *AI Expert,* April 1990, p. 72.

11. Layden, John E., and Thomas A. Pearson. "A Missing Link in Total Quality." *Controls & Systems,* March 1992, pp. 42–44.

12. Montague, G. A., M. T. Tham, M. J. Willis, and A. J. Morris. "Predicative Control of Distillation Columns Using Dynamic Neural Networks." Research paper from the Department of Chemical and Process Engineering, University of Newcastle-upon-Tyre, Newcastle, United Kingdom.

13. NeuSIGHT product literature. Mentor, OH: Pegasus Technologies Corporation, 1994.

14. Parker, Kevin. "Advanced Stats for Manufacturing Beyond SPC." *Manufacturing Systems,* June 1993, pp. 32–37.

15. Parker, Kevin. "Lending Expertise to Supervisory Control." *Manufacturing Systems,* August 1993, pp. 24–28.

16. Pinto, James J. "Process Industries—Trends in the Nineties." *Automation,* September 1990, pp. 40–41.

17. Process Insight product literature. Austin, TX: Pavilion Technologies, Inc.

18. Rowan, Duncan A. "On-line Expert Systems in Process Industries." *AI Expert,* August 1989, pp. 30–38.

19. Rytoft, Claes, Bo Johansson, Kent Bladh, and Nicholas Hoggard. "Aspects of Future Control Systems." *Control Engineering,* September 1990, pp. 135–38.

20. Sprow, Eugene E. "What Hath Taguchi Wrought?" *Manufacturing Engineering,* April 1992, pp. 57–60.

21. Tinham, Brian. "Control in the Chemical Industry." *Control and Instrumentation,* January 1993, pp. 34–35.

22. VerDuin, William H. "Neural Nets for Predictive Monitoring and

Control of a Phosphate Coating Line." IPS '91 Proceedings of the 20th Annual ESD International Programmable Controllers Conference, ESD—The Engineering Society, Detroit, Michigan, 1991, pp. 471–79.

23. VerDuin, William H. "Optimizing Combustion with Integrated Neural Networks and AI Technologies." *Control Engineering,* July 1992, pp. 38–40.

24. Warren John. "Model-Based Control of Catalytic Cracking. *Control and Instrumentation,* July 1992, pp. 33–34.

Chapter Nine

Technology Trends:
Aspects of Future Computer-Based Systems

9.1 INTRODUCTION

The previous chapters described some history of design and manufacturing technologies and the technologies and applications of knowledge-based systems in these areas. This history, and the 20–20 hindsight it provides, will help system developers repeat past successes and learn from past failures. The history is also useful because it points to many compelling but unpursued opportunities. But knowledge-based technologies, and computer technologies in general, continue to evolve. This evolution will provide new opportunities by overcoming previous limitations.

This chapter contains a view of the future or at least of trends in knowledge-based systems and computer-based technologies as they relate to applications in design, manufacturing, and management. These trends are presented to help the reader not only anticipate the future but more specifically to chart implementation strategies and tactics that will best meet evolving needs and most effectively leverage future opportunities.

The trends in this chapter include changes in technologies and application environments and needs. Trends in knowledge-based system products are then described.

9.2 NEEDS AND OPPORTUNITIES

9.2.1 Shrinking Staffs, New Tasks

Skilled employees will continue to pursue other opportunities, and so expertise will be lost. Those called upon to design or produce products or run the businesses that do these things will find themselves short of required experience or problem-solving skills.

Worldwide competition will continue unabated. To remain players in this new marketplace, manufacturers must radically restructure the way they do business. An element of this change is that new technologies, new products, and new markets impose ever-increasing requirements for the acquisition and interpretation of information. These two trends are negative in that they force change in the ways that design and manufacturing are done. These trends can be positive in the sense that new information technologies enable organizations to seize a competitive advantage from better use of information.

9.2.2 Computer Price Wars

Another bad news/good news story is that of computers. From the point of view of computer manufacturers, the competition is brutal, the pace is frantic. Computers get faster and cheaper. Profit margins are dropping, and surveys such as a recent study by management consultants McKinsey and Co. conclude that half of today's top computer vendors will no longer be top competitors by the year 2000.

From the perspective of computer buyers, the computer hardware story is mostly good news. As prices drop and performance climbs, previously unimaginable power is available on users' desks. A Digital Equipment Corporation advertisement in the August 17, 1993, *Wall Street Journal* describes Digital's latest high-end PC (personal computer) as "powerful as the original $7.5 million CRAY-1 supercomputer." Very sophisticated software tools can now be supported by inexpensive computers. This provides new application opportunities, as lower prices make justification easier.

For some users, there will be bad news in the form of obsolete hardware and maintenance that will become more expensive and more difficult to obtain. This will be true for owners of discontinued computer models from vendors still in the hardware business, and particularly for owners of hardware from vendors no longer supplying hardware.

Since computers other than PCs use proprietary, or vendor-specific, operating systems, software for obsolete hardware may also become obsolete. Users will be forced to either replace both hardware and software or pay increasingly high prices to maintain existing software. Because of these software and hardware maintenance issues, buyers should give thought to the long-term prospects of a computer vendor before committing to a purchase.

These statements about hardware and software support do not apply to PC purchases. This is not due to an absence of competition in the PC market. It is ruthless; and a shakeout of PC hardware vendors can be expected. However, buyers need not be overly concerned because their investments in PC hardware and software are at minimal risk due to a unique feature of this market.

The PC market is huge and is served by many hardware and software vendors. This is because new vendors are free to enter this market, and by competing with existing vendors, they drive prices down. The reductions in price and, to a lesser degree increases in performance, have driven the market growth.

This is a unique phenomenon in the computer business and can be directly attributed to IBM, which launched the huge PC market by creating the first open (nonproprietary) operating system. By publishing the details of the PC's operating system DOS (Disk Operating System), IBM enabled any vendor to build a machine that was functionally if not physically identical to IBM's, in that it could run any of the many DOS-based software packages. Again, bad news/good news: IBM is challenged by the presence in the marketplace of many low-priced *clones,* but users benefit by the availability of inexpensive computers.

On the software side, the proliferation of PCs created both the opportunity and the desire for software developers to create DOS-based products. Users have benefited by the growth of a market that has generated a wide variety of powerful yet inexpensive software. One can debate the technical merits of DOS and the limitations that it imposes on designers and users of software, but the price/performance attributes of this unique market make these limitations largely irrelevant.

Returning to the topic of a PC hardware vendor shake-out, the failure of clone manufacturers will limit users' maintenance options on these machines. However, owners of these PCs will find that many individual circuit boards within these DOS-based computers are interchangeable, and so their orphan machine may remain usable for years. As PC prices continue to drop, users can also afford to replace their older machines.

They will be increasingly motivated to do so by the trend of PC software developers to assume the presence of a fast processor (CPU) and plenty of RAM and hard-disk memory in the user's PC. Since this is increasingly true, software developers take advantage of hardware-based opportunities to provide the best performance and most features for what is becoming the largest audience. The only downside is that owners of obsolete PCs will find their software choices increasingly limited.

9.2.3 Open Systems

The growth of the PC market is a positive trend, indeed, as PCs transform engineering, manufacturing, and management. Interestingly enough, Japan has only recently initiated a comparable open system approach. As a result, the PC market in Japan has been tiny, on a per-capita basis, and is only now experiencing the explosive growth that the open system approach has provided U.S. users.

The benefits that open systems have provided PC buyers have not been lost on buyers of workstation hardware and software. Buyers want to know why workstation software can't be more like PC software, that is, inexpensive and able to run on any machine. As PCs become more powerful, the performance advantage of low-end workstations over PCs is eroding. The result is that workstation vendors are increasingly in competition with high-end PCs.

This has created two trends. First, low-end workstation software and other software are becoming less expensive, in some cases equaling prices of high-end PC products.

The distinction between PCs and workstations is blurring in other ways. Customers want wider choices in both hardware and software by being able to run software from any vendor on hardware from any other vendor. Thus, customers are pressing workstation vendors to support an open systems approach.

VMS, a traditional proprietary operating system from Digital Equipment Corporation, is now presented as an open systems product called Open VMS. The UNIX operating system has always been promoted as an open system. However, the reality has been that each vendor has used a unique variation of UNIX. The variations may be similar, but regardless, vendor X's software wouldn't run on vendor Y's hardware. The trend, driven by customer insistence and a credible alternative in the form of powerful PC hardware and operating systems such as NT and OS/2, is toward a truly open version of UNIX. Lower workstation hardware and software prices are anticipated as a result, providing good news for users and bad news for vendors.

9.2.4 Software Design

Diminished staffing and budgets are driving hardware and software trends. Buyers of computer-based systems increasingly require low acquisition and implementation costs. This has led to a trend toward more application-specific software. Vendors respond to increasingly segmented markets. This trend is apparent in knowledge-based systems, but also in software of all types. Users benefit because less customization and therefore less staff time are required to implement software that is already set up for specific tasks and environments.

Smaller staffs and budgets also drive the trend toward more efficient software development and maintenance. Software development is now one of the least automated tasks, but new technologies and programming approaches are emerging to streamline this development task. This trend will benefit not only commercial software developers but also those charged with software maintenance.

Maintenance may be performed by the software developer or by the user. Either way, it is a difficult (and thankless) task. The difficulty arises in trying to understand the relationship between the problem to be solved and the software that does so. Software reflects the detailed thought processes of the original system designer and programmer. It is difficult to recreate these thought processes to understand how they are reflected in code.

It is particularly difficult to discover how various sections of code are related. In real life, documentation may be cryptic and may not fully describe how sections of code are linked. If these ways are not apparent to any one except the software's author, it is not hard to see why software modifications often have unintended consequences (also known as *bugs*).

Software design practices are changing to meet development and maintenance needs. The traditional software design practice, called *structured programming,* involves a main program loop that is executed again and again, each time calling upon various subroutines, each of which contains specific functions. Data are transferred from subroutine to subroutine but not managed by any one entity. The shared access to data can create problems if one subroutine inadvertently corrupts data used by another subroutine.

Object-oriented programming addresses this problem and thus provides simplified software maintenance by explicitly managing access to data. The subroutines of structured programming are replaced by objects. Each object incorporates one or more functions and also incorporates and controls the associated data. Objects communicate with each other by messages flowing through interfaces that transfer information to and from objects but shield internal procedures and data from the outside world.

This encapsulation prevents unanticipated interactions between various procedures and database elements. This clustering of software functions also streamlines documentation and maintenance. New functions can be added by adding an object. Existing functions can be modified by revising an object.

Software objects often correspond to real-world objects. For example, one software object might be a motor controller, with inputs consisting of start and stop commands, and outputs including overtemperature and overcurrent indications.

Changes in user needs also impose new requirements. Staffing levels on the factory floor are often cut, and in many cases, problem-solving (not to mention reading) skills are decreasing. These trends are shaping the design of manufacturing support systems so that they require less time and skill to use and less up-front training. User interfaces reflect this trend. Presentation of data and graphs that help users analyze problems is being replaced by systems that perform the analysis and tell the user what to do, or at least what the choices are.

Knowledge-based systems will play an increasingly vital role in this pursuit of easier-to-use systems. Embedded expertise and analytical capabilities increasingly enable systems to interpret data and reach conclusions. The incorporation of knowledge-based technology is increasingly evident in both design and manufacturing systems. Also increasingly evident in these systems is the way that a knowledge-based component of a system is seamlessly integrated and thus indistinguishable from the traditional system it supports.

9.2.5 Agile Manufacturing

Worldwide competition is driving another trend. Many manufacturers have found the need to establish a local presence in marketplaces worldwide. Local manufacturing operations may be required to achieve access to markets or to reduce tariffs. Local marketing operations will be in closer touch with consumers and better able to respond to changes in needs.

The trend toward worldwide operations will be reflected in the increasing scope of manufacturing and management support tools. As manufacturers buy raw materials and make and sell products in international marketplaces, they will increasingly rely on tools that support multinational, multicurrency operations. They will also require tools that support distributed decision making, as far-flung operations function as semi-autonomous units to best meet the needs of rapidly changing marketplaces.

These changes in the ways of doing business require more than new manufacturing and management tools. They require a new mindset that recognizes that the pace of change will not diminish and that ongoing success will require continued receptiveness to new needs and rapid response to those needs. This concept is captured by the phrase *agile manufacturing*.

Advanced manufacturing technologies such as **robotics, flexible manufacturing systems (FMS),** and **computer-integrated manufacturing (CIM)** have focused on rapid response on the manufacturing floor. Agile manufacturing requires more. The point is that not just manufacturing but all elements of an organization must become agile to respond quickly when required. The organizations that will succeed in an international and interconnected economy are those that are not only competitive in all respects but can also respond rapidly when windows of opportunity open.

Agility may be required to respond to rapidly emerging market opportunities before the competition. These opportunities may involve new products, new processes, or new markets. Agility may

also be required to successfully overcome new constraints or leverage new partnering opportunities.

Each of these opportunities may require rapid and substantial
changes in ways of doing business to enable this new manufacturing
and management perspective to be put in place. Unwavering top
management support will be necessary to accomplish the required
change in corporate culture that will be required to transform a
traditional organization into an agile one.

The challenge will be greatest within large organizations. Large
organizations enjoy economies of scale and staffing levels that provide ready access to specialists and support staff, but the downside
has traditionally been slow response to changing needs. The success
of small companies has been their innate agility. One of the few
benefits of small size is that there is minimal bureaucratic infrastruc-
ture to filter inputs and delay outputs. The challenge will be to
combine the responsiveness of a small organization and the skills
and facilities of a large one.

Computer technologies will evolve to meet the needs of agile
manufacturing. New systems will be the logical evolution of existing
technologies such as CIM. The premise of CIM is that computer-
based systems supporting various activities in an organization
would be seamlessly integrated. This often-used term will take on
new meaning as integration becomes more seamless and as the
scope of integration broadens.

Manufacturing and management support tools will integrate not
only across applications such as design, testing, and manufacturing,
but also across layers within an organization, across physical locations, and even across industries to link suppliers, subcontractors,
manufacturers, and their customers.

Emerging software tools, increasingly incorporating knowledge-
based technologies, will support agility by helping existing staff
acquire the knowledge and analytical capabilities to reconfigure
products and processes. New tools will help staff establish new
marketing, support, and investment programs to meet rapidly
changing needs. Enabling existing staff members to acquire new
skills and become effective in a faster-moving environment will
be a challenging task. But it will be increasingly supported by
knowledge-based systems in decision-support and training roles.

With new training and support systems and new design and manufacturing technologies, there will still be opportunities that call for resources that cannot be provided in house or justified on a long-term basis. For this reason, an increasingly common approach will be to overcome limitations by the formation of strategic partnerships. Companies will pool strengths and together pursue new opportunities.

This approach is well established in marketing and distribution, with manufacturers' representatives and independent distributors a common sales channel. Partnerships in design and manufacturing are now emerging in the electronics and automobile industries. As an example, Nissan and Ford jointly designed the Mercury Villager/ Nissan Quest minivan, with Nissan dominating the design phase and Ford having sole assembly responsibility at their Avon Lake, Ohio, plant. These arrangements conserve resources and provide the partners opportunities they would not enjoy on their own.

9.2.6 The Virtual Corporation

The logical extension of this trend is to form partnerships to meet current needs and to dissolve these partnerships when those needs have been met. Members would then form new partnerships to address new opportunities. Over time, organizations would expect to be joined in ongoing but continually changing partnerships.

Each member of such a partnership would provide a resource necessary to meet a common goal. The resource might be needed skills, facilities, or access to markets or capital. Together, the members of what might be called a *virtual corporation* would provide resources that meet the needs of a new opportunity, but without the overhead costs and inertia that are typically required to provide such capabilities within a single organization. Costs would be lower and response quicker.

The challenge will be to make this new type of partnership work. There are management issues to be addressed, such as how to make negotiation of contractual and financial arrangements simple enough to respond quickly to new needs. There are also significant issues in communications and information management, such as ensuring that opportunities are identified in a timely manner and that collaboration is managed in a mutually beneficial way. But trends in competitive needs and emerging communications and computer technologies are expected to make the virtual corporation a reality.

The trend in communications is clear. Enhanced communications is a common theme in recent new product successes such as VCRs, portable and cellular phones, and fax machines. The expanding capacity and merging of computer, telephone, and cable TV networks is creating national information superhighways.

New products and services will emerge to fill these superhighways. Some will be new home entertainment offerings such as 500 channel cable services, and others will build upon *smart* home information systems consisting of an integrated TV and computer.

Other new products and services will serve purely commercial purposes. New forms of information and new ways of providing that information will extend current information management opportunities and enable new opportunities such as virtual corporations.

This information will be diverse. CAD drawings and associated information might be generated in one organization and modified in another. The parts described in these drawings might be built in a third organization, assembled in a fourth, with billing and shipping documentation generated in a fifth.

The information will support other activities and may include marketing information, unstructured requests for quotations, and hot tips. It may include product specifications, pricing, and shipping information. It may include contracts and other legal documents. (IBM is reported to be taking a leadership role in this area by considering electronic format, that is, computer-based, documents to be legally binding in the absence of paper copies of those documents.)

The linking of organizations in new and changing partnerships will create the need to share knowledge at higher levels. Managers will require more diverse knowledge, abstracted and summarized to meet their more unstructured problem-solving and issue-identification needs. Knowledge-based systems will play a critical role. They will track issues and trends in the worldwide marketplace to identify new opportunities for current and yet-to-be formed partnerships. They will help provide the framework and the detail to support these partnerships.

9.2.7 Real-Time Systems

Another trend is driven by the need for more timely information. Timeliness is an element of the agility that is increasingly required to respond to new opportunities and to better manage existing operations. New communications products and services will transfer information faster. But changing expectations in the timeliness of information are also creating the need for new ways to process information.

Traditionally, information has been available after the fact, describing events that have already occurred. Accounting functions are inherently after-the-fact since their role is to report on prior events. But in other applications, historical information is used not because it is appropriate or preferred, but because it has been the best that was available.

Statistical process control reports on prior results, with a time lag introduced by the need for the operator to collect and analyze data. Manufacturing operations data might be collected, processed, and reported on a daily, weekly, or even monthly schedule. By the time the report is received, a problem may have intensified or a large quantity of defective product built. In addition, it may be difficult to reconstruct, so long after the fact, the circumstances that led up to the problem.

Both enabling technologies and user needs are driving a trend toward real-time systems, systems in which data are collected and analyzed as fast as they are generated. More timely reports minimize deviations from desired conditions and help problem solvers link observed effects to underlying causes. The benefits are clear in process monitoring and control applications, in terms of better quality, reduced scrap, and reduced costs. In general management applications, benefits include all the ways that faster response can benefit the organization, from faster response to new problems and opportunities to better control of costs.

Another trend is to move beyond real time into predictions of the immediate future. Better modeling technologies will enable easy development of more comprehensive and accurate forecasts. As forecasting becomes more useful, it will be increasingly incorporated in manufacturing and management decision support systems. In manufacturing applications, the ability to predict changes in system state would enable control systems to anticipate deviations and implement preemptive control actions. In management applications, better forecasting might help overcome fluctuations in foreign currency exchange rates and raw material pricing and availability. Better forecasting may provide more useful market projections and help managers plan optimal strategies for purchasing, production, and marketing in light of these projections.

The trends described above will be reflected in future computer hardware and software and in the use of these products. These trends are also evident in knowledge-based technology products as described below.

9.3 PRODUCTS

Knowledge-based technology products reflect the trend toward application-specific tools. Designers of these tools are developing more focused knowledge-based technologies that are narrower in scope but better suited to specific applications. This benefits users who can focus on the problem to be solved, not the details of the technology being employed. In general, less system configuration is required, and performance is better. Examples of this trend are described in the following section.

9.3.1 Character Recognition

General-purpose neural net software and hardware is widely available and has been successfully used for applications such as image recognition. A new application-specific neural net hardware/software product has demonstrated significant performance benefits for its targeted application. A design focused on the specific requirements of magnetic ink character recognition (MICR) points the way to future applications in manufacturing.

A cashier who is asked to accept a bank check wants to be sure that the check is not drawn on a high-risk account. This is done by first reading the magnetic ink characters that identify account and transit number on the bottom of the check. This account information will then enable a telephone-based credit risk ranking service to advise the cashier to accept or reject the check.

The traditional MICR reading approach is to manually key in the numbers, a slow and error-prone method. Electronic MICR readers are available but require a second pass of the check through the reader about 20 percent of the time. VeriFone has developed a neural net-based MICR system that is reported to be less expensive and more accurate than traditional MICR devices. The neural MICR is claimed to be 99.6 percent accurate, even with wrinkled, folded, or marked checks.

This performance is enabled by a special Synaptics (San Jose, California) neural net chip. This chip employs an inexpensive design with minimal power consumption, both of which help make the system practical for point-of-sale applications. The MICR device includes an infrared light-emitting diode (LED) that illuminates the characters to be read. The reflected image is transmitted through a lens to a retina on the neural net chip. Combining the camera and electronics on one chip simplified manufacturing and provided a more direct and therefore faster path between the camera and the neural net.

The low cost and high performance design of this particular chip may lend itself to manufacturing applications. The broader concept of designing an integrated and specialized system of knowledge-based technology, electronics, and peripheral devices to meet a narrow range of applications can be expected to be widely copied.

9.3.2 Predictive Control

Another new product category is that of advanced controls. Change comes slowly to the control industry because customers are conservative and, in particular, reluctant to accept new risks. But as competitive pressures require better performance from all parts of an organization, manufacturers are increasingly motivated to pursue higher performance control strategies.

Processes can be candidates for higher performance control for two reasons. First, process characteristics may change appreciably over time due to changes in equipment, raw material, or operating environment. Second, process dynamics, that is, the response of the process to changes in controlled or uncontrolled parameters, may not be accurately represented by the relatively simple mathematics of traditional controls. In both cases, the result is that a process runs more slowly or with more deviation from desired setpoints than is possible with more sophisticated strategies.

One way to achieve better performance is through model-based control. A model of a process is developed and used to predict changes in the process. Control action requirements are anticipated on the basis of this model. In this way, control actions are taken before the process has deviated to a significant degree from the intended state. This not only minimizes deviations from optimal performance but also reduces the risk of the system moving irretrievably far from the desired state. It is on this basis that it is possible to run the process faster and thus more effectively.

Model-based control has been implemented in critical applications in which the considerable effort involved in model development and maintenance was justified by considerable improvements in process capabilities or cost. But less critical applications, or those for which a major development expense could not be justified, were controlled by less-expensive but less-effective methods.

Neural nets are redefining model-based control opportunities. Neural net-based models streamline the model development and maintenance task and, by reducing implementation cost, expand application opportunities. The next step is to provide high-performance neural net approaches that move beyond real-time control to model-based predictive control. The resulting systems will combine easy implementation with true model-based control capabilities. Manufacturers will then be able to achieve the better performance of model-based control on a wider variety of applications.

9.3.3 Manufacturing Support

Manufacturing support is another application area in which new capabilities and a new product category are emerging. This new category is called by a variety of names, among them **manufacturing execution systems (MES).** It can be viewed as the logical extension of MRP (material requirements planning) and MRPII (manufacturing resources planning), making it in essence MRP III.

MRP helps manufacturers determine raw material and component requirements to support a given production schedule. MRPII is a broader-based planning tool that includes manufacturing facilities along with materials as resources to be scheduled. MES continues in this direction by recognizing the need to integrate manufacturing planning with planning for the organization as a whole. The goal is to provide software tools that support the inherent interactions between manufacturing, design, purchasing, marketing, finance, and distribution.

Emerging MES products provide a framework to integrate new and existing tools for each of these areas. They address the need to integrate sources of information and to support analysis and decision making in physically dispersed and perhaps autonomous organizations.

An emerging extension of the MES concept might be called an *intelligent interface*. The goal of this interface would be to provide a higher-level integration of tasks and tools. This interface would incorporate advanced database and knowledge-based technologies to provide an information management layer on top of existing databases and software tools.

The advanced database technologies in the intelligent interface would address the need to provide easier access to information. These technologies would provide more convenient access to data and problem-solving tools that are distributed across a computer network. To seamlessly integrate software, tools, and data, the intelligent interface would, of course, provide a single user interface to all of the databases and tools. But the interface would also learn from experience which databases and records within those databases a user calls upon. The intelligent interface would automatically generate the arcane commands traditionally required to access data in large relational databases, freeing the user from the need to remember and use these commands.

The intelligent interface would also contain knowledge-based technology to interpret data. This capability would match the trend in factory-floor software away from presentation of data toward interpretation of data and presentation of issues and actions. By discovering relationships in data, neural nets would identify trends that bridge various functions in the organization. Expert systems might incorporate knowledge on analysis to be performed and actions to be taken in response to these trends.

The result will be a system that is more useful as a management tool for managers. Its scope will be broader, across the entire organization. Its functionality will support interpretation as well as acquisition of knowledge. Details of data access will be automated, and so the system will depend less on specialized skills. In these ways, an intelligent interface on top of an MES system will be more supportive of the unstructured, constantly changing tasks of the manager.

To summarize technology trends, the computer business is going through changes that will present users some bad news and mostly good news. Knowledge-based technologies will play an increasing role, supporting the trend away from presentation and toward interpretation of data.

Technology trends will be an element in overall business trends. Competition will remain fierce, and the successful organizations will be those that respond rapidly and effectively to new requirements and new opportunities. Information management will become an increasingly critical competitive factor in a worldwide marketplace. Knowledge-based systems will thus play an increasingly visible and important role.

The challenge is to translate opportunity into action. The technology and products trends in this chapter are intended to provide a perspective that may be useful in identifying not only specific knowledge-based system applications, but also longer-term information management strategies. The next logical step is to act upon this insight, and the next chapter provides some guidelines on how to do this. The focus is on lessons learned. This final chapter covers the steps of identifying, justifying, planning, and implementing knowledge-based systems.

9.4 REFERENCES

1. Babb, Michael. "Neural Nets: No Hype, Please." *Control Engineering,* June 1993, p. 49.
2. Babb, Michael. "What DCS Operators Want." *Control Engineering,* November 1991, p. 53.
3. Bhatt, Siddharth. "The Control Connection." *Chemical Engineering,* May 1992, pp. 91–94.
4. Brule, Michael, and Robert Chronister. "Computers: The Next Decade." *Chemical Engineering,* August 1993, pp. 74–84.

5. Donlin, Mike, and Jeffrey Child. "Is Neural Computing the Key to Artificial Intelligence?" *Computer Design*, October 1992, pp. 87–99.
6. Hammerstrom, Dan. "Neural Networks at Work." *IEEE Spectrum*, June 1993, pp. 26–32.
7. Inglesby, Tom. "Manufacturing Software Heads for the 21st Century." *Manufacturing Systems*, January 1993, pp. 18–32.
8. Keyes, Jessica. "AI on a Chip." *AI Expert*, April 1991, pp. 33–38.
9. Lavey, Tom. "Manufacturing Software: Looking beyond MRPII." *Manufacturing Systems*, May 1990, pp. 62–66.
10. McKaskill, Tom. "Applying the Process Production Model to Discrete Manufacturing." *Manufacturing Systems*, October 1992, pp. 70, 72.
11. Metz, Sandy. "Making Manufacturing Better, Not Just Faster." *Managing Automation*, August 1990, pp. 22–24.
12. Meyer, Fred. "Object-Based Systems Do Windows Better." IN-TECH, July 1993, pp. 39–41.
13. Robinson, Mike. "Beyond EDA." *Electronic Business*, June 1993, pp. 42–48.
14. Rytoft, Claes, Bo Johansson, Kent Bladh, and Nicholas Hoggard. "Aspects of Future Process Control Systems." *Control Engineering*, September 1990, pp. 135–38.
15. VerDuin, William H. "Neural Nets: Software That Learns by Example." *Computer-Aided Engineering*, January 1990, pp. 62–66.
16. *The Wall Street Journal*, August 17, 1993, p. A11.
17. Young, Lewis H. "Pessimism Invades the Computer Outlook." *Electronic Business*, March 1993, p. 128.

Chapter Ten

Getting Started:
Identifying Opportunities and Implementing Projects in the Real World

10.1 INTRODUCTION

Having read about technologies, applications, and trends, the question is, what's next? Hopefully, the answer is to identify and pursue knowledge-based systems applications within the reader's organization. The available tools and the application history suggest that the elements for successful applications exist within most organizations. The needs are there. In most cases, the resources can be made available if the project is correctly identified, justified, planned, and executed.

Each of these project management steps is important. Incorporating a new technology such as knowledge-based systems into the plan requires extra care in some steps. Management will be justifiably curious as to whether the new technology has merit in their organization and if the conditions for successful application can be met. Those planning the implementation should be prepared to answer questions up front, and ultimately deliver a successful system.

This chapter is a summary, in a sense, of the previous chapters. The background in knowledge-based technologies and the application examples are designed to help readers identify the best applications and reduce the risk of new applications.

This chapter contains planning issues, application issues, criteria for success, and action items. Such information may be useful as the basis for a check list to be prepared as a first step in identifying, justifying, planning, and implementing a solution. The underlying concept is that a successful project must be both justifiable and feasible and that these criteria must be clearly established up front. This chapter is designed to enable the project planner/manager to do this.

Technical and general management will also be interested in these issues. They will want to understand knowledge-based technologies and their significance. They also need to understand application issues and benefits to better convince themselves and persuade others of the viability of a proposed project. This will be necessary if for no other reason than that it will be their responsibility to provide the resources and personal support to ensure successful development and integration of the system into the business.

The following sections consist of action items presented in roughly chronological order to accomplish the tasks of identifying and justifying an opportunity and planning and implementing a knowledge-based system.

10.2 IDENTIFY OPPORTUNITIES

10.2.1 Identify Business Issues

All organizations confront issues regarding their ongoing viability. It may be the competitive environment or regulatory or legal issues. It may be the ability to attract and retain the scarce expertise that distinguishes one organization from another in its ability to rapidly and effectively respond to new needs. It may be the ability to successfully apply new product or process technologies.

Most organizations confront a variety of issues. The point of this task is not to create a laundry list of issues but rather to identify one or two fundamental issues that affect, or soon will affect, the organization. In terms of justifying a knowledge-based system, these issues should relate to the capture, organization, or use of knowledge. As the previously described applications indicate, knowledge-based systems can address virtually all areas of manufacturing and management.

Identifying these types of issues may be unfamiliar territory and perhaps not of great interest to technical staff members. But the incentive to become knowledgeable, if not interested, in this area is that answers to this and the next few questions will be vital in securing management interest. Management's willingness to devote resources and accept some level of risk will be directly related to the degree to which a given solution addresses a real problem. A technology in search of a mission will be a tough sell. But selling a technology-based solution to an acknowledged problem will be far easier. The first step is to identify that problem.

10.2.2 Identify Business Needs

The next step is to identify what changes are required in the way an organization does business. The answer at this point is probably not any particular new technology but instead may be the need to respond to fundamental changes in the business environment, such as constraints in the production of by-products or the inability to use critical raw materials. Or an organization may need to address increased competition with designs that provide more features, better quality, or lower prices. Faster delivery or faster response to new requirements may also be required.

To a degree all of these requirements are true for every organization. A business need argument becomes more compelling when it is as specific as possible. It should refer to needs of the reader's organization, in the context of characteristics of the industry in which it competes and the strengths and weaknesses of present and future competition. These needs will tend to be defined in strategic terms. The logistics that will meet these needs will be identified later in the project-planning process.

10.2.3 Identify Business Opportunities

In this step, a particular need is restated as an opportunity in disguise. If the main competitive disadvantage of an organization is the quality of its products, then an opportunity exists to gain (or at least retain) market share on the basis of improved quality. If competitive products are of lower cost, an opportunity exists to gain market share on the basis of better design or manufacturing techniques to lower product lifetime costs, including production costs, as well as preproduction costs of design, redesign, and testing, and after-the-sale costs such as warranty or recall costs.

Several opportunities may be linked. Many organizations confront competition with more modern design and manufacturing methods that offer both lower costs and better quality. In these common cases, one general need for, say, better design methods may translate into a number of related opportunities. With luck, the opportunities are positive, and gains in market share or profitability can be anticipated. But not to be overlooked are the opportunities to avoid bad news: By investing in design or manufacturing tools, a manufacturer can retain market share or avoid regulatory costs.

10.2.4 Identify Design and Manufacturing Issues

Business opportunities can, of course, affect all areas of an organization. However, since this book deals with design and manufacturing, focus will be maintained, and the next step is to identify those design and manufacturing issues that relate to previously identified business needs and opportunities.

A design issue, for example, might be that several redesigns are necessary to meet product specifications and provide manufacturing feasibility, and so costs and timing for design, prototyping, and testing are excessive.

A manufacturing issue might be the upcoming regulation of a process by-product or the inability to maintain acceptable product quality with existing staff and equipment.

10.2.5 Identify Design and Manufacturing Needs

This step brings into focus the specific needs in design or manufacturing capabilities that must be addressed to resolve the previously identified business needs.

A design-related need might be to ensure faster and uniformly high-quality designs by implementing better design procedures that integrate knowledge in the manner of concurrent engineering and capture best practices in each area.

A manufacturing need might be to lower direct production costs by 5 percent or scrap levels by 10 percent. Or a more fundamental need may exist to put in place a completely new manufacturing process and, with that process, new operating and maintenance skills.

10.2.6 Identify Design and Manufacturing Opportunities

In this step, it is time to translate general needs into a specific technology-based opportunity. A design or manufacturing opportunity is identified, as is the knowledge-based system that will realize that opportunity. The proposed solution will be identified by both application area and enabling technology. For example, the proposed solution might be a neural net-based process optimizer to minimize consumption of energy and generation of an undesirable by-product. Or an expert system may be required to capture scarce design expertise and that system combined with other tools such as a materials database and a neural net-based material selection tool.

The proposed solution should be defined as the solution to a clearly identified design or manufacturing problem, which in turn is related to a specific business need. By having worked through many steps of defining business and technical needs, the proposed solution can be realistically evaluated on the basis of its ability to meet the full range of needs. This analysis minimizes technical risk and identifies benefits of the proposed solution.

10.2.7 Do a Reality Test

Before investing time in developing a detailed project proposal, do a reality test to be sure that the basic requirements for success can be met. Be sure that

a. The stated needs and business opportunities are acknowledged by management, or at least can be made visible and relevant.
b. The proposed solution convincingly addresses the stated needs and opportunities.
c. The use of knowledge-based technologies can be shown to provide clear benefits.

d. The required type and depth of knowledge for the proposed system is available or can be acquired.

e. Management support can be anticipated or generated.

f. Realistic expectations and ongoing support from users and management can be anticipated.

Regarding the first three items, only the reader and perhaps his or her management can provide answers on business needs and opportunities. Management's interest in pursuing technology-related opportunities depends on corporate history and culture and how well a project is justified.

Regarding item (d), the previous discussion of knowledge-based technologies and applications provides insight on knowledge requirements and overall technical feasibility, once the other conditions have been met. A plan will be required to either capture existing knowledge or acquire it from outside sources. It is worth thinking through this point early on, because it will affect project feasibility, benefits, and resource requirements for both development and maintenance.

Items (e) and (f) may be difficult to know in advance. Positive indicators would be a successful design or manufacturing system implementation in the recent past and a general willingness to address issues and implement solutions. The converse is also true. If the failed project was high-tech, it may dissuade management from considering another advanced-technology project, even if the specific technologies are different. However, these skeletons in the closet are not insurmountable. Many organizations have successfully implemented a second system once an appropriate cooling-off period has elapsed since an unsuccessful first try.

Some managers are perpetually in search of the *silver bullet*—the new technology that, alone, will solve all of the organization's problems. This perspective is helpful in selling applications of new technology, especially ones such as knowledge-based technologies for which such sweeping claims have been made. But a project justified on this basis runs the risk of unrealistic expectations. No knowledge-based system alone can overcome problems such as obsolete products or manufacturing facilities.

Realistic expectations will enable a technology and project to succeed on its own merits, and not, for example, deem a design

optimization project unsuccessful because it alone could not re-
vive a long-lost market. Thus, realistic expectations about what a
knowledge-based system will and won't do will be helpful in implementing a second knowledge-based system (or any other new technology).

10.3 JUSTIFY THE PROJECT

10.3.1 Define the Benefits

The background work from previous steps provides the basis for
the argument. It is now time to define why management should
contemplate spending money and perhaps disrupting operations, if
only briefly, to implement a knowledge-based system. The case
must be persuasively made that the proposed project will bring
clear benefits outweighing costs.

In some organizations, new technologies are actively embraced.
These decision makers understand that one common feature among
organizations that succeed over the long term is that they continually evolve in the way they do business and that one element of
this evolution is the timely introduction of appropriate new technologies. These organizations tend to take a proactive stance on new
technologies. Many of these firms have ongoing programs to track
technological trends, with tracking done by staff members explicitly charged with identifying and implementing new technologies.

In such organizations, knowledge-based systems tend to be an
easy sell, and historically the majority of knowledge-based systems
have been implemented in such organizations. However, the proven
successes and greater maturity of knowledge-based technologies is
enabling this approach to be sold into more conservative organizations on the same basis as other improvements in facilities, tools,
or methods. In one case, a knowledge-based quality improvement system was sold as a quality system, not as a knowledge-based or even
computer-based tool. This was done to enlist the support of management interested in quality systems but not too interested in computers.

Selling a project on the basis of results requires that benefits be
defined in nontechnical terms. Benefits might be stated qualita-
tively. One might say, for example, that the decreasing market
share trend will be reversed.

Ideally, the benefits can be expressed as specific measurable improvements, and these numbers can be justified. For example, an argument might be made that, based on similar experiences elsewhere, a knowledge-based system will enable product costs to be cut by X percent and that this in turn will provide Y percent greater sales and Z percent greater revenues. This argument would need to include the basis on which the current situation is assumed to be analogous to the previous one.

Decisions to provide new design or manufacturing tools are often made in the same way that decisions to improve manufacturing facilities or any other investment decisions are made. In return for a certain amount of money to be invested, benefits will be received in the form of greater revenues or lower costs. The investment decision is made on the basis of the financial attractiveness of what may be viewed as nothing more than a business transaction.

The merits of an investment can be stated in a number of ways. Most organizations have a preferred method, and project justifications should be presented in that manner. Two common approaches are **payback period** and **return on investment (ROI).**

Payback period is simply the time period over which increased revenues or decreased costs will pay back the initial investment. For example, added revenues or reduced costs of $10,000 per month would pay back a $100,000 project in 10 months.

Each company will have its own maximum acceptable payback period. This time period may change, depending on economic conditions and, thus, management's tolerance for risk and interest in investment proposals. Many companies require paybacks of 12 to 24 months, although the argument can definitely be made that adequate preparation for the future requires more patience and a longer-term view.

Return on investment (ROI), also known as internal rate of return (IRR), is expressed as a percentage and represents the investment yield in the form of cost savings or revenue increases that will result from investment in the project. ROI calculations are somewhat more complicated than payback period and are described in detail in accounting textbooks.

Briefly, the calculation involves the time value of money, as would be involved, for example, in the calculation of mortgage payments for a given principal and interest rate. Banks provide mortgages—in essence renting money—because they get back enough additional money through interest that it is worth waiting 15 or 30 years to get it all back. The same concept applies to technology investment decisions. A dollar spent today will be returned many times over in the future.

The assumptions used in ROI calculations vary from company to company. Those individuals justifying projects on this basis will want to determine their organization's calculation methods as well as what is considered an acceptable rate of return.

The time invested in generating the investment payback numbers and doing the calculations will be repaid by a justification process that is easier because it is based on (reasonably) hard numbers rather than on faith. Justifications based on qualitative benefits may be compelling, but they may suffer in the competition for scarce investment dollars. Nonetheless, there are many cases in which investments were made basically on faith, and that decision was clearly justified in retrospect. Many advanced technology implementations have been in this category. Such investment proposals are difficult, but not necessarily impossible to sell.

As in any sales encounter, a key to effective selling is to discover the customer's *hot button,* the topic or perspective of particular interest. In this case, the benefits story should be designed to intrigue the manager from whom funding approval or other support is sought. It might be worthwhile to learn before preparing the project justification about any perspectives on the part of the funding manager that might relate to the way the justification can be most effectively presented. The justification will also, of course, call upon the homework done in earlier tasks to identify needs and opportunities.

10.3.2 Justify the Approach

The next step is to describe why the suggested approach is the best way to achieve the stated objectives. The proposal needs to contain the appropriate level of detail, striking a balance between a high level of detail that may overwhelm and confuse—and perhaps distract from the underlying benefits story—and a sketchy outline that may raise questions about the depth of understanding and thoroughness of planning.

The project description should include the application type and location such as a design optimization system for the corporate headquarters design group. The description should also specify the technology to be used, the products and services involved, and the staffing requirements. Staffing requirements would include the internal experts and anticipated users required for system development and validation, as well as for user training and system maintenance. Outside services might be required to assist in tool selection and system design, to provide domain expertise, and perhaps to provide complete system development, installation, and user training.

The newness of knowledge-based technologies and the incidence of overblown claims may cause managers approached for funding and support to ask difficult questions. The person proposing the projects should be ready with answers based on research into technologies and applications and input from prospective suppliers of knowledge-based tools and services as required. These answers will generate confidence that the project has been carefully and critically evaluated, that anticipated product vendors offer robust, maintainable tools appropriate to the task at hand, and that services requirements and sources have been identified.

Products can be identified by reviewing advertisements in trade magazines, including those from the application area of interest, as well as those focused on knowledge-based technologies. Magazines such as *PC AI* and *AI Expert* publish lists of vendors with short product descriptions.

Project planners should request marketing and application information from vendors. Candidate products should be screened with potentially tough questions such as:

- What applications have you or your customers done in my application area?
- Is the application currently in use?
- Whom may I talk to about it?
- How long have you been in business? (to judge likelihood of future product or custom system support—a potentially significant issue in an industry with considerable turnover of vendors.)
- What product support do you offer (bug fixes, upgrades, telephone support).

- What services do you offer (installation and training services, product extensions to suit special requirements, custom system development and support).

The answers to these questions will help the project planner select sources of tools and services and will also help clarify approaches, risks, and risk containment strategies. All of this information will be helpful in justifying, planning, and implementing the project.

Assuming that the justification is successful, the next step is to plan the project.

10.4 PLAN THE PROJECT

10.4.1 Knowledge Issues

Planning a knowledge-based system is like planning any other project: think it through. Know that there will be surprises. In knowledge-based projects, the biggest surprises tend to be in acquiring, interpreting, and validating knowledge. These steps are harder or take longer than anticipated because

- Experts are not available.
- Experts disagree.
- Data is sparse or difficult to acquire.

Other surprises have to do with computers in general:

- Available computers are not suitable (they are too slow, have too little memory, or they don't support graphics).
- Computers, networks, databases, operating systems, or software packages are incompatible.

The knowledge-related issues are generally present to some degree. The successful knowledge-based systems work around these problems by finding alternative sources of knowledge and perhaps spending more time on system validation to ensure that the resulting system has not been compromised.

10.4.2 Hardware and Software Issues

The computer-related issues are becoming less of a factor. As computers become faster and cheaper, the likelihood of access to an acceptable machine increases. Knowledge-based system tools are also becoming available on a wider variety of computer platforms, with high-end PCs becoming a more common choice to replace workstations. However, the project planner may want to verify that, for example, users will have access to graphics display capabilities if such an interface is planned.

Some knowledge-based systems are completely self-contained, incorporating data within the application or at least in the same computer. This neat, clean approach avoids many compatibility issues. However, there are also knowledge-based systems that call upon external data. This might be done for functional or maintenance reasons. This common reliance on external data has clear benefits but also imposes implementation challenges.

Some knowledge-based systems call upon real-time data. The benefit of this external data source is obvious: timely access to data enables rapid analysis and corrective action. From an implementation perspective, access to these data may require extra hardware. The computer in which the knowledge-based system resides will need a communications card and a cable to connect it to the process control hardware into which the sensors are connected. A communication module may also need to be added to the process controller.

Other knowledge-based systems call upon static data located in external databases. This may be preferred to bringing the data to the same computer running the knowledge-based system if the database is too large to move or copy. In addition, leaving the database and its owners and maintainers intact has the significant benefit of ensuring that the database is properly maintained without added effort.

Once the decision is made to use external data sources, the project planner must ensure this access. The increasing use of organization wide computer networks helps in this regard. Proprietary databases, proprietary operating systems, incompatible file formats, and incompatible networks can all conspire to make communication difficult. But the trend is in the direction of standard formats, and so (perhaps slowly) this problem will diminish.

The project planner should consult with the appropriate computer system manager, who will most likely be aware of such issues and may also be able to identify solutions. In some cases, software or hardware bridges may be required to link incompatible formats. Commerical bridges are becoming increasingly common, and so a custom solution may not be required. In other cases, the implementation of such bridges may not be justified, and arrangements can be made to transfer data on a regular basis from a remote database to the knowledge-based system's self-contained database. This approach bypasses compatibility issues at some expense in ongoing maintenance.

10.4.3 Project Planning

Other elements of project planning are like any other technology or application area. The planner should outline

a. Project scope.
b. Measures of success.
c. Project milestones that will be used to track progress, identify roadblocks as they occur, and revise strategy or reallocate resources as required.
d. Required interaction with vendors, users, domain experts, and others.
e. Internal and external staffing requirements for system development, validation, installation, training, and maintenance.
f. *Critical path* items, that is tasks for which delays will either slow or stop project progress.
g. Caliber and intended uses of documentation.
h. Clearly defined deliverables.
i. System maintenance plan and responsibility.

The reader's organization may use standard forms and formats for this outline. Regardless of format, this information will be important in the planning, implementation, and after-the-fact analysis of the project.

10.4.4 Products and Services Issues

If commercial software will be involved, the planner should carefully select the tool to meet project requirements, understand what is included in the price, and what support services—including maintenance—are available.

If outside software development staff are to be employed, the planner should check their track record in the area of interest. The planner should require a detailed proposal covering the elements for which they will be responsible. This proposal should outline all of the above project-planning items unless there will be shared development responsibility.

Shared responsibility can impose significant management challenges. Two teams must be kept in contact to keep abreast of each other's activities and to work to common goals. It is also important to have a clear understanding at the outset of a project on the responsibilities of each team and the boundaries of those responsibilities. This may be useful if problems arise during development to determine ownership of the problem.

A clear understanding of responsibilities and tasks will also be useful in securing accurate and complete proposals for development services. The outside developers will look to the project planner to provide a clear picture of project scope, available knowledge resources, and other details. This will enable the outside developer to provide a so-called *fixed-price* or *firm fixed-price* quotation that is fair to both the buyer and seller of development services.

A fixed-price contract puts the risk of project estimation on the outside developer, since the commitment is made to provide certain deliverables for a certain fee, regardless of the actual level of effort required. The outside developer should have a comprehensive understanding of requirements to avoid having to either pad the estimate or risk underestimating the job and losing money on it.

If the items that determine project magnitude, such as development scope and availability of knowledge are not known up front, an alternative that may be considered is to enter into a time-and-materials contract. This less structured arrangement enables the outside developer to bill for services and materials as required, typically to reach a specified goal.

Since, in this case, the risk of project estimation is now on the project planner, he or she must clearly define how specific tasks and other requirements will be established and how progress will be measured. The planners should also define how time/cost overruns will be handled for individual tasks and whether a not-to-exceed limit should be set. Since this arrangement eliminates the need to accurately estimate the level of effort required for a given project, outside developers will prefer this arrangement. Buyers of services are often understandably reluctant to use this approach.

Variations on the fixed-price (FP) and time-and-materials themes are possible, particularly in contracts with the U.S. government. The time-and-materials concept is embodied in *cost plus fixed fee (CPFF)* contracts. These contracts cover time and materials costs, and to these a fixed fee is added. CPFF contracts are also often limited by a not-to-exceed dollar amount.

To limit the buyer's risk in nonfixed-price contracts, performance incentives may be provided. These incentives motivate the seller to control time-and-materials costs by providing fees based on performance. One approach is the *cost plus incentive fee (CPIF)*. In this arrangement, the fee is negotiated as a percent of the target cost, but the difference between the target and actual cost will vary the fee according to an agreed-upon formula. Under this arrangement, it is possible for a seller to receive a negative fee, that is, a loss if costs climb too high.

Another incentive arrangement is known as *fixed-price incentive fee (FPIF)*. This contract calls for a maximum total cost. If actual costs are below this number, the seller is provided a fee that varies according to an agreed-upon formula in which reductions in actual costs are rewarded by greater fee percentages.

Nongovernment buyers of services may not encounter these many types of contracts. However, they should be aware of the need to define tasks and minimize contractual as well as technical risk as much as possible.

10.4.5 Intellectual Property Issues

Project planners must also reach agreement with outside developers on ownership rights for both the software to be developed for hire and the intellectual property, that is, data and knowledge, that will be incorporated in the system. Some issues to be considered include the following:

* The project planner should ensure that his or her organization, as the buyer of custom (as well as commercial) software, will not be at risk from patent infringement or other misdeeds or legal entanglements, either intentional or unintentional, by the outside developer.

- The software may be used as required. The developer may require that software licenses be purchased for each location or even each computer in which the custom software is installed. Other developers provide unlimited usage licenses or full ownership rights to the purchaser of custom software. There is no particular right answer here other than to be sure that the buyer's needs will be met.
- The project planner should ensure that both parties' proprietary information will be held in confidence, typically for a limited period of time such as 3 to 10 years.

Custom software developers will want separate licensing and ownership arrangements for any of the developer's proprietary technology and software incorporated in the custom system. In this case the buyer would receive, perhaps for a fee, a right-to-use license for the required proprietary technology but would have no other rights to it.

The importance placed on knowledge-based systems as a competitive advantage is reflected in not-uncommon security arrangements. The existence of a project, or even a relationship between a buyer and a developer of knowledge-based systems, may be considered proprietary information. The buyer of custom software may choose (typically for a fee) to prevent access by competitors to the software or even the general knowledge or calculation methods within it. These arrangements also typically incorporate a time limit.

Custom system buyers may anticipate an up-front payment to initiate the project. Subsequent payments might be tied to completion of milestone tasks and approval by the buyer's designated representative (usually the project planner). Short-term, inexpensive projects (say, under $50,000) might involve only an initial and final payment, with the latter payable when the software is demonstrated and documentation delivered according to the terms of the contract.

The above details and more would typically be covered in some combination of a formal proposal, a letter contract, a software license, and a **nondisclosure agreement.** Once the paper shuffling is done and approvals obtained, it is time to begin the project.

10.5 IMPLEMENT THE PROJECT

10.5.1 Project Management

Often, the project planner becomes the project manager. The manager should begin the project with a meeting between members of the system development team and intended users. Development and maintenance roles will be discussed at this meeting.

Developers should immediately start cultivating interest and input from the intended users. Their input will be vital to ensure that the finished system truly meets user needs. Their enthusiastic initial support will also help ensure that the system is actively embraced when delivered. These acceptance and maintenance issues must be addressed early on, because these two elements are critical for project success.

The project manager must also continue to enforce realistic goals and provide access to the resources necessary to meet those goals. Even with sincere upper management support, the domain experts and other staff members most valuable in system development can be diverted from support of the knowledge-based system because they are often the staff members most useful in supporting day-to-day operations.

The project manager should monitor all aspects of knowledge acquisition. The manager must ensure that available knowledge resources are sufficient to accomplish the agreed-upon objectives. The nature of these resources depends on the particular knowledge-based technology.

For expert systems development, the manager might want to determine the following points:

- Are the experts cooperative and forthcoming?
- Can consensus be reached among multiple experts?
- Do the rules cover all aspects of the agreed-upon problem domain?
- Are the answers and the explanation of the answers correct?

Fuzzy-logic project questions might include:

- Are control objectives clearly stated?
- Does a consensus exist on control objectives?

- Does the control strategy suit the control objectives and accommodate available sensor data, control hardware, and controlled system hardware?

Neural net questions might include:

- Do available data sufficiently characterize the system? (That is, will the neural net develop a complete and accurate model, or are more data required?)
- Does the trained net replicate actual system behavior?

It is important to know early in the project whether these questions can be answered affirmatively. This will enable new resources to be acquired or system scope to be modified. To make this judgment, the project manager must incorporate periodic reality tests in the project plan. These tests will be designed to measure completeness and accuracy of acquired knowledge against some predefined expectations. This can be a difficult measurement to define. The system developer comes to know the problem domain and the relevant issues only as the project progresses. In the absence of this understanding, it is difficult to know up front the total scope of required knowledge in order to estimate and test its acquisition rate.

One solution is to test knowledge acquisition results directly by developing a series of system prototypes. These prototypes are presented to the harshest and most knowledgeable critics available—the intended users of the final system. This will reveal gaps and errors in system knowledge. It will also highlight structural and philosophical differences between the ways that experts view and solve problems and the ways that the knowledge-based system attacks these same problems.

For a knowledge-based system to be embraced by intended users, it must meet their needs for problem-solving style as well as substance. The multiple prototype approach is effective because it helps developers identify changes required not only in completeness and accuracy, but also in overall perspective and problem-solving approach.

For this reason, it may not be reasonable to assume that development of the final system consists of incremental enhancements to an evolving prototype. In fact, several prototypes may be scrapped during the course of the project. This does not diminish the value of these discarded prototypes as knowledge acquisition and system design tools. It instead reflects the common situation that problems that are complex enough to justify a knowledge-based system also require a fair amount of work to fully understand.

The project manager undertaking demonstration of prototype systems must be sure to set the stage for these reviews. Final system users who will review a functionally and visually incomplete prototype might be inclined to dismiss the entire development process and prejudge the final system on the basis of that early prototype. They must understand that the prototype they are about to see is a work in progress and not indicative of the final product. Reviewers must know that the developers know that prototype is incomplete and perhaps off the track, but that the final system will become good only through the input and support of the reviewers. This ongoing attention to knowledge and system verification will help ensure that the final system will meet agreed-upon objectives and user needs.

During the course of development, the manager should be thinking of additional knowledge-based system opportunities. These opportunities may lie in problems that are extensions of the problem addressed by the system under development, or they may be problems of a similar nature in other areas. In either case, the project manager should be aware of these additional opportunities, so that he or she can be ready to define and justify new projects as the first project nears completion. The justification may be easier because the first system provides a useful foundation for the second system. In any case, the existence of one successful system will validate the approach and greatly simplify justification of the second.

10.5.2 Planning Ahead

Getting started in knowledge-based systems is much like starting any project that involves a new technology or a new way of doing business. There are issues to be addressed related to the technology and other issues related to changes in ways of doing business. A knowledge-based system project also has much in common with any project, in that it should be carefully planned and monitored, with clearly defined resources and expectations.

Past successes in knowledge-based system development and improved computers and knowledge-based tools suggest that many more successful knowledge-based systems will be developed. Developing and installing a knowledge-based system is not only useful, it's fun. There is an intellectual challenge in bringing together business needs, technical needs, and the opportunities presented by technology and knowledge. There is a sense of accomplishment in putting together a system that may be the first of its kind in the organization and that users embrace as an interesting and useful tool.

Implementers will be well positioned as the management of information becomes an increasingly visible opportunity and knowledge-based systems become integral parts of successful organizations. The individuals capable of identifying opportunities and delivering solutions will enjoy increasing opportunities as organizations seek out these new problem-solving skills.

The relatively short history of knowledge-based systems has shown that the approach is useful, that clear-cut benefits can be realized, and that, with realistic expectations and proper planning, knowledge-based technologies can deliver cost-effective solutions not practical by other methods. This opportunity to innovate and solve problems is not to be missed.

10.6 REFERENCES

1. Houghteling, James L., Jr., and George G. Pierce. *The Legal Environment of Business*. New York: Harcourt, Brace & World, 1963.

2. Kerzner, Harold. *Project Management: A Systems Approach to Planning, Scheduling, and Controlling*. New York: Van Nostrand Reinhold Company, 1979.

3. Keyes, Jessica. "Why Expert Systems Fail." *AI Expert,* November 1989, pp. 50–53.

4. Vancil, Richard F., ed. *Financial Executive's Handbook*. Homewood, IL: Dow Jones-Irwin, Inc., 1970.

Technical References

The focus of this book is primarily on applications of knowledge-based systems. Note that the references included at the end of each chapter are primarily application-oriented and support the applications in that chapter. Some readers may be interested in a more rigorous technical treatment of knowledge-based system technologies and related topics. For this reason, the following references are provided.

For readers interested in more depth in *expert systems:*

1. Cohen, Paul R. and Feigenbaum, Edward A., eds. *The Handbook of Artificial Intelligence* (4 volumes). Stanford, CA: Heuris Tech Press, and Los Altos, CA: William Kaufmann, Inc., 1982.
2. Rich, Elaine. *Artificial Intelligence.* New York: McGraw-Hill, Inc., 1983.
3. Winston, Patrick Henry. *Artificial Intelligence,* 2nd ed. Reading, MA: Addison-Wesley Publishing Company, 1984.

For readers interested in more depth in *fuzzy logic* and *neural networks:*

4. Bezdek, James C., and Sankar K. Pal. *Fuzzy Models for Pattern Recognition-Methods That Search for Structures in Data.* New York: IEEE Press, 1992.
5. Kosko, B. *Neural Networks and Fuzzy Systems: A Dynamical Systems Approach to Machine Intelligence.* Englewood Cliffs, NJ: Prentice Hall, 1988.
6. Pao, Yoh-Han. *Adaptive Pattern Recognition and Neural Networks.* Reading, MA: Addison-Wesley, 1989.
7. Simpson, Patrick K. *Artificial Neural Systems Foundations, Paradigms, Applications, and Implementations.* New York: Pergamon Press, 1990.
8. Zadeh, L. "Fuzzy Sets." *Information and Control.* 8 (1965), pp. 338–53.

For readers interested in *fundamental mathematics* underlying knowledge-based systems and technology underlying *computer vision systems* and *parallel computer hardware architectures* (as might be used to implement a high-speed neural network):

9. Dougherty, Edward R., and Charles R. Giardina. *Mathematical Methods for Artificial Intelligence and Autonomous Systems.* Englewood Cliffs, NJ: Prentice Hall, Inc., 1988.

Readers may also be interested in the journals of the Institute of Electrical and Electronic Engineers, Inc., New York, NY—in particular:

10. *IEEE Spectrum* magazine.
11. *IEEE Transactions on Computers,* a publication of the IEEE Computer Society.
12. *IEEE Transactions on Neural Nets,* a publication of the IEEE Neural Networks Council.
13. *IEEE Transactions on Pattern Analysis and Machine Intelligence,* a publication of the IEEE Computer Society.
14. *IEEE Transactions on Systems, Man, and Cybernetics,* a publication of the IEEE Systems, Man, and Cybernetics Society.

Glossary

The following glossary is intended to provide a brief overview of some of the terms used in this book that relate to knowledge-based technologies, and design and manufacturing applications of these technologies. This list is by no means exhaustive in the area of knowledge-based technologies, nor does it incorporate terminology specific to the applications described. The reader may also want to refer to the index for further detail on the following terms and to better understand the context of their usage.

adaptive A computer-based system that adapts or modifies its behavior to suit changing circumstances. In the context of process or machine control, it is a control system that is made aware of changes in the process or machine's operating characteristics or external environment and is able to change the control system characteristics to compensate for these operational or environmental changes. This is an important characteristic since few manufacturers have staff available to continually monitor operational or environmental changes and adjust control system parameters to suit. Adaptive control can provide more effective control that better meets objectives without additional operator or maintenance staff involvement.

agile enterprise An organization that can rapidly and effectively respond to changes in customer requirements, operating conditions, and partnering opportunities. This management perspective, believed to be critical to long-term manufacturing success, will be enabled by new management and organization structures, new computer-based tools, and a heightened reliance on the use of information.

AI *See* **artificial intelligence.**

algorithm A procedure, often a set of equations, for solving a problem or accomplishing a specific result.

ANN Artificial neural net or neural net. *See* **neural net.**

architecture 1. In neural net technology, architecture refers to the arrangement of the neurons and interconnection among neurons and the way that the neural net learns to self-improve its own performance. There are dozens of different types of architectures, each with specific strengths and limitations. Within each general type of architecture, users can choose the size of the net, that is, the number and arrangement of neurons and interconnections. 2. In computer technology in general, architecture refers to the physical and logical design of the computer, including the central processing unit (CPU). The evolution of CPU design is driving the increases in computer speed and is therefore an area of interest from a marketing as well as technical perspective.

artificial intelligence A branch of computer technology that aims to extend computer capabilities into areas in which humans have been effective but computers have not. Examples include reasoning, pattern recognition, learning, and self-improvement. Knowledge-based systems are a branch of artificial intelligence (AI).

artificial neural net *See* **neural net**

ASCII (American Standard Code for Information Interchange) An industry standard character code that is used to transfer information between databases and between computers. As a character code, it represents a way to translate letters, numbers, and symbols into binary data that can be digitally transmitted. Many commercial databases and spreadsheets generate ASCII files.

backward chaining A reasoning method used in expert systems. An outcome is assumed, often consisting of a current situation. The expert system starts from this conclusion and uses rules within its rule base to step backward in time. The result is an analysis that (presumably) recreates the steps leading up to the present situation. The alternative approach is forward chaining, in which an initial condition is assumed and rules are used to predict a future outcome.

bang-bang control A simple control approach in which a device is either fully on or completely off or, in the case of motion control, moved to one end of its travel or the other. More sophisticated approaches allow some gradation of control between the two extremes to provide partially on or intermediate positioning.

batch process A production process for which a characteristic weight or volume can be associated, on the basis that the weight or volume of product is processed as a single entity. The alternative is a continuous process in which the product is made on an ongoing basis with no segregation into batches.

best match A data-retrieval function of the CAD/Chem product design tool that finds the closest match to, for example, a desired product recipe or product characteristic. This function is not a data look-up, because the match need not be exact.

by-product An undesired or unanticipated product of a process. For example, slag is a by-product of the steelmaking process. It must be disposed of and, as a marginally useful material, represents one of the costs of making steel.

CAD *See* **computer-aided design.**

CAM *See* **computer-aided manufacturing.**

chaining *See* **backward chaining** and **forward chaining.**

CIM *See* **computer-integrated manufacturing.**

classify Assigning a new data point or circumstances to a particular predefined group, upon the recognition of similarity between the new data point or circumstance and the characteristics of existing members of that group. The classification capability is the ability to determine similarity. The act of classification is a useful problem-solving step because the identification of membership in a particular group may further clarify the problem or suggest a method of analysis or solution.

closed-loop control Enabling the controller of a process to automatically vary its own settings in response to measured deviations, or error, between the desired and the actual value of a process output parameter. The alternative, open-loop control, requires action by an engineer or operator to change process setpoints as required.

CMM *See* coordinate measuring machine.

CNC *See* computer numerical control.

computer-aided design (CAD) Computer-based tools that, at a minimum, replace the drafting task of product design. CAD continues to evolve, with increasing analytical and modeling capabilities that help designers calculate the functional requirements of a new product and assess the appearance and performance of a proposed design.

computer-aided manufacturing (CAM) Computer-based tools that support manufacturing tasks such as planning, scheduling, and production. Production support may include direct implementation of CAD designs by automatically configuring flexible production equipment such as robots or **flexible manufacturing systems (FMS)** to suit the CAD design requirements. The trend towards standard data formats will make the transfer of files from CAD to CAM systems easier and more common. This and the increasing use of flexible production equipment will make integrated CAD/CAM systems common.

computer-integrated manufacturing (CIM) The use of computers to link activities and sources of data throughout a manufacturing organization. This information management strategy is aimed at better decision-making based on faster access to a wider selection of data. Organizationwide computer networks help implement this strategy. CIM trends include products that overlay individual software tools and databases to provide a uniform user interface and to help integrate the tools and databases under the common interface.

computer numerical control (CNC) Computer-based numerical control in which the data manipulation, logic, and control capabilities of computers extend the capabilities of numerical control to replace manual and mechanical process setups with automatic and electronic setups based on stored dimensional and other process data.

computer vision A system combining an electronic camera, a computer, and special hardware and software to acquire and interpret images. These images may be snapshots of production parts, and the computer vision task is to identify the type or orientation of the part. Or the task may be inspection for the presence or location of features or to verify color, texture, or surface finish. Practical issues include positioning and lighting the part beneath the camera, but typically the more difficult issues are those of interpreting the image rapidly and accurately, in a way that captures features of interest without being distracted by irrelevant differences.

concurrent engineering A management strategy that seeks to minimize needless manufacturing complexity and start-up problems. New product design is treated as a concurrent, or collaborative, task between the design and manufacturing engineers. By acquiring practical input regarding manufacturing capabilities and constraints early in the design processes, potential manufacturing problems can be resolved easily and inexpensively.

confidence interval A measurement of data quality, based on statistical methods, that addresses the issue of uncertainty. Errors in measurement ensure that no experimental data point is exactly correct. Calculations based on those data are also subject to that error. The confidence interval is the range of values within which a given parameter is expected to fall with a certain probability or level of confidence. One can be more confident that a number is somewhere within a wide range, or less confident that it is in a narrower range, because one can be sure that no data point can be known exactly. Error can be minimized but not eliminated by better experimental methods and larger sample sizes.

constraints In design applications, limits in the use of materials, design features, product capabilities, or costs; in manufacturing applications, limits in process capabilities, scheduling, availability of raw materials, generation of by-products, or costs. The utility of knowledge-based systems is based in part on their ability to provide answers that reflect these constraints.

consult A neural net problem-solving functionality. A neural net discovers relationships in data that may be used as a problem-solving model. For example, a relationship may be discovered between the design and the performance of a product. In this case, the neural net is trained to develop a design model. The neural net is then consulted to provide, in this case, a new product design and, in the general case, classification or estimation on the basis of the learned relationship.

continuous improvement A management approach that addresses the need to meet new requirements in product features, quality, and cost. The concept of continuous improvement is that all members of an organization are empowered to become agents in identifying and solving problems and that these solutions will provide continuing incremental improvements in manufacturing that will be reflected in continually improving products.

continuous process A production process that is run on an ongoing basis. The process is shut down or modified to change product characteristics as required. But unlike a batch process, there is no characteristic volume or weight of product associated with the process.

controlled parameters Those process parameters that are controlled to meet process objectives. Other parameters, such as ambient temperature and humidity, may affect the process but are not controlled. To best meet process objectives, the controlled parameters may need to be controlled in a way that compensates for changes in uncontrolled parameters.

coordinate measuring machine (CMM) A device that measures dimensions of parts with a scope ranging from verification of critical dimensions up to generation of a complete three-dimensional representation of a part. Some CMMs employ a hard-tipped stylus to locate points on the part surface; others detect reflected laser light beams.

correlation A statistical measurement of the degree to which a change in one parameter is reflected in a change in another. Positive correlation indicates that the two parameters change magnitude in the same direction. Negative correlation indicates that one decreases in magnitude as the other increases.

crisp A fuzzy-logic term that describes a set whose limits are precisely defined. An example of a crisp set is those temperatures above 68° and below 72°. A fuzzy set, in contrast, has less clearly defined boundaries. An example of a fuzzy set is those temperatures at which most people feel comfortable.

critical path A project-planning term that refers to the sequence of tasks that, if delayed, will delay the entire project. Some tasks are independent of others and of sufficiently short duration that they represent essentially no risk in terms of delaying the project. Tasks composing a critical path are linked, in that one must be completed before another can start. If any of these tasks is delayed, the overall project would be delayed by a like amount.

data analysis A data preprocessing functionality of the CAD/Chem product design tool that identifies duplicate or conflicting data or variables that are unchanging, linearly dependent, or weakly correlated with another variable. All of these problems should be corrected before these data are used to develop a neural net- (or statistics-) based model.

DCS *See* **distributed control system.**

decision tree The logical structure of an expert system in which the answer to one rule presents one or more additional questions to be answered, but eliminates others from consideration. The structure can be diagrammed in the form of a tree, and each answered rule moves the system further up to smaller branches representing narrower, more specific rules.

defuzzification The output functionality of a fuzzy logic system. The system accepts imprecise data as input and manipulates those data according to fuzzy rules. The results of this process are translated into a precise, or crisp, output according to one of several defuzzification processes.

design of experiments (DOE) A method to define experimental requirements. DOE will identify the range of experimental conditions required to support the development of a model valid over a given range of values. DOE is particularly useful, that is, the recommendations are not obvious, when designing experiments involving a large number of independent variables.

desirability function An optimization goal definition functionality of the CAD/Chem product design tool. The premise is that multiple-objective optimization must reflect user-defined ranking of objectives. The desirability function describes the relative desirability of different parameter values across the allowable range of that parameter. It is used in conjunction with an overall ranking of that parameter's importance.

discrete part A product whose output can be counted rather than measured by weight or volume. Discrete parts range from fasteners to electronic and mechanical components to assemblies of these components.

disk operating system (DOS) The operating system for IBM and IBM-clone personal computers (PCs). DOS is the most popular PC operating system, with the operating system for Apple Macintosh computers a distant second. The operating system manages the internal operation of the computer, handling input and output functions, memory management, and the execution of programs. Operating system issues from a user's perspective include user friendliness and software compatibility. A major strength of DOS is the variety of software available as a direct result of the popularity of DOS-based computers. Microsoft's Windows and NT, and IBM's OS/2 operating systems have evolved from DOS.

distributed control system (DCS) A process control hardware scheme common in batch and continuous process industries. Control capabilities reside in control units located on each piece of process equipment close to the location of sensors and devices connected to the controller. This distributed arrangement contrasts with the more centralized programmable logic controller (PLC) approach more common in discrete part manufacturing.

DOE *See* **design of experiments.**

domain expert A person expert in a particular problem-solving domain. This expert would be called upon to provide analysis and problem-solving advice that would be incorporated in a knowledge-based system. A common approach is to capture this expertise in the form of rules that, in turn, constitute the rule base of an expert system.

DOS *See* **disk operating system.**

EAM *See* **episodal associative memory.**

embedded 1. In the context of technology, embedded refers to one technology used in conjunction with, and in a supporting role to, another technology in a way that the embedded technology is not visible to the user. 2. In the context of systems, embedded refers to one system, such as a knowledge-based system, that is self-contained but used in support of and embedded within another system in a way that the internal knowledge-based system is not visible to the user of, for example, a smart CAD system.

episodal associative memory (EAM) A neural net-based technology that captures design and manufacturing expertise as episodes. This streamlines knowledge acquisition and extends neural net capabilities into additional application areas.

estimate A neural net-based functionality in which the decision support model developed by the neural net is used to predict, or estimate, the response to given situation. For example, a design model might define the relationship between product design and product performance. For a hypothetical new design, the neural net-based model might estimate new product performance characteristics.

expert system A knowledge-based system that incorporates problem-solving knowledge in the form of rules. These rules are typically acquired from a domain expert—a person skilled in the analysis and resolution of a particular class of problem. Alternatively, rules may be acquired from handbooks and other sources of information. The expert system user initiates a dialog with the system, providing answers to structured questions posed by the system. In this way, the system successively narrows the range of possible problems down to a single problem and then presents the answer to that problem.

expert system shell A software development tool that supports the development of an expert system. The shell contains a user interface and an *inference engine,* the logic-processing mechanism that manipulates rules within the system. The use of a shell enables the system developer to focus on acquiring knowledge and composing rules specific to the problem of interest, with minimal development required to provide generic expert system capabilities.

faceted part representation A three-dimensional representation of a part used to define that part for a rapid prototyping process. The surface of the part is represented as a continuous surface composed of discrete three-sided facets. Each facet shares edges with adjacent facets and has an associated *surface normal vector,* a vector that describes whether the facet is on an inside or outside surface of the part.

feature-based design An advanced CAD (computer-aided design) capability. Design features such as holes and slots can be identified by the user, enabling the user to manipulate these higher-level features rather than always working with each of the details, such as individual dimensions and tolerances, that make up each feature. Feature-based design is thus a time-saving and knowledge-capturing feature.

fire In the context of expert system or fuzzy-logic rules, a rule is *fired,* or deemed to be true and acted upon, if the conditions for that rule are met.

first principle A basic law of chemistry, physics, or other natural science. First principles express fundamental theory and are used as the basis for decision-support models, as are statistical methods and, now, neural networks.

flexible manufacturing system (FMS) A manufacturing system, typically a machining center, that can be electronically programmed. Part designs from a CAD (computer-aided design) system provide electronic input to the FMS and define the part to be produced. Manual setup is avoided, and so FMS is uniquely capable in small lot size manufacturing. Production rates are limited by the single part/single process format.

FMS *See* **flexible manufacturing system.**

formulation The recipe, or proportions of ingredients, for many products made by batch or continuous processes. The design challenge is to find the formulation that best meets product performance, processing, and cost requirements. One manufacturing challenge is to monitor and control the formulation and other aspects of the process. Another challenge is to optimize the process to accommodate changes in raw material and equipment characteristics and to best meet multiple objectives.

forward chaining A reasoning method used in expert systems. An initial condition, such as the current situation, is assumed. The expert system starts from this point, successively applying rules within its rule base to step forward in time. The result is a predicted outcome. The alternative approach is **backward chaining,** in which an outcome is assumed and rules are used to step backward in time. In this case, the result is a predicted cause or starting point.

fractionization The process of breaking down complex hydrocarbons into their various constituents. For example, crude oil is fractionized into gasoline, kerosene, fuel oil, lubricating oil, and asphalt.

fuzzy logic A knowledge-based technology focused on control applications. Control objectives are stated in fuzzy, or qualitative, rules. This avoids the need for precise, or crisp, rules as required by expert systems. Fuzzy logic requires fewer rules and is more tolerant of sensor data errors than expert systems. Fuzzy rules require greater computation time than expert system rules, and so specialized hardware is often required.

fuzzy rule The type of rule characteristic of fuzzy logic and used to state control objectives. A fuzzy rule defines a qualitative relationship that may be expressed as a smoothly varying fuzzy set membership function. This contrasts with a crisp, either-or membership function that is assumed in the use of other technologies.

graphical user interface (GUI) A style of computer display design. This advanced style of user interface screen replaces text and simple graphics with *icons,* or pictorial representations of specialized tasks, as well as common tasks such as "delete files." The user can initiate and monitor several tasks at once through windows that provide visual interaction with each task. The several common formats of GUI are distinguished by their look and feel—that is, their overall appearance and methods of identifying and accessing common functions.

GUI *See* **graphical user interface.**

heat rate In a combustion process, the rate at which heat is generated. In electric power generation, for a given level of electrical output, one objective is to increase efficiency by minimizing the heat rate, or use of fuel. Other objectives may include minimizing emissions and providing smooth transitions from one power output level to another.

heuristic In expert systems, a rule of thumb or logical jump. Sometimes an expert can bypass the normal small steps involved in identifying a problem by evaluating and eliminating alternatives. An expert familiar with a particular type of problem may recognize, for example, that there is one possible underlying cause that dominates the other possible causes. This heuristic, or insight, may be incorporated in an expert system that would jump to investigation of the most likely cause before starting a slower and more methodical search.

hidden layer An element of a neural net architecture. The neurons in a neural net may be in an **input layer** (*see also*) and thus receive input to the net, or may be in an **output layer** (*see also*) and provide the output function of the net. The neural net may include one or more layers of neurons in so-called hidden layers between the input and output layers. Depending on the type of neural net, multiple hidden layers may be required to enable the neural net to fully capture the complexity of a relationship to be learned. Too many hidden layers and too many neurons can also create modeling errors, making the relationship appear more complex than it is.

IF-THEN A common type of expert system rule describing IF a particular situation exists, THEN the following question should be asked, analysis performed, or action taken.

input layer An element of a neural net architecture. The neurons in the input layer of a neural net provide the input function of the net. The input layer neurons may be connected to neurons in the **hidden layer** or layers (*see also*) or to the **output layer** (*see also*) if no hidden layer is used. The input layer must include at least as many neurons as inputs, since each neuron accepts one input. Neural net architectures such as the Functional Link Net allow the user to define additional inputs made up of combinations and transformations of original inputs. This enables more complex relationships to be captured without the computational complexity imposed by the alternative of more hidden layers.

integrated product/process design (IPPD) An engineering management approach in which the design of a product and the associated process are undertaken concurrently. Product and process designs are most effectively integrated by knowledge-based and other computer-based tools that help designers identify interactions between product and process designs and resolve these interactions in ways that best meet overall objectives. The IPPD process identifies production problems earlier in the design cycle and helps provide lower cost, more effective solutions to these problems.

intellectual property The value of a product other than its physical embodiment. In software, as in printed material, it is the research, development, composition, insight, and creativity that provide the user value beyond that derived from the physical floppy disk or book. The protection of intellectual property is a critical issue for software developers, because it is the result of expensive and time-consuming development and is the basis of their business. Software users may also encounter intellectual property issues relating to the expertise or data they incorporate in software written by themselves or others.

intercorrelation Synonymous with correlation. *See* **correlation.**

investment casting A metal working method, in particular, a specialized method of casting. Casting is typically used to create complex geometries and is considered a near-net-shape method because the cast part requires relatively little machining to acquire its final features and dimensions. Investment casting is used for particularly precise and complex parts. A wax mold is made in the shape of the desired part. A ceramic shell is formed around the wax model. The wax is melted out of the shell, leaving a thin ceramic shell that is used as the mold for the final metal part.

IPPD *See* **integrated product/process design**

ISO 9000 A European quality control and auditing procedure. The ISO standards writing organization divides this overall methodology of defining, implementing, and monitoring compliance with quality control procedures into numbered elements such as ISO 9001, and 9002. Compliance with these procedures is becoming a more common requirement for sales into the European market.

iterative A synonym for *repeated*. In the context of calculation methods, it refers to a process of changing a parameter slightly, rerunning a calculation, changing that parameter again (and typically incrementally further in the same direction), rerunning the calculation again, and so on until a solution is found or an objective reached.

JIT *See* **just in time.**

just in time (JIT) An inventory management method in which buffer inventory is greatly reduced or eliminated. Traditionally, mismatches between process usage rates and relatively infrequent raw material and component deliveries were handled by the storage of supplies of these materials equal to perhaps weeks or months of production. As an element in streamlining production and cutting inventory holding and facilities costs, manufacturers have implemented JIT scheduling of material and components delivery. Shipments from a given supplier may arrive several times per day, in the order in which they will be used. Material handling is thus simplified, once the specialized material handling and computer-based ordering and scheduling systems are in place and the new manufacturing and management systems debugged.

knowledge-based In the context of computer software technologies, the incorporation of advanced analytical or problem-solving capabilities. The term knowledge-based denotes a technology that extends computer capabilities beyond those of traditional record keeping and calculation. Knowledge-based systems may incorporate diagnostic and problem-solving experience, or abilities to discover relationships, and use these discovered relationships to solve design and manufacturing problems, among others.

knowledge engineer An engineer who acquires knowledge and learns its structure and application so that it can be most effectively reused to solve new problems. The task of knowledge engineering is most often associated with the manipulation of symbolic knowledge by expert systems.

layer An element of neural net architecture referring to the arrangement of neurons in the net. Layers may be characterized as **input layers, output layers,** and **hidden layers** (*see also*), each with a specific role. The arrangement and number of layers vary among different neural net architectures. The choice of architecture and number of layers and neurons within each layer determines the capabilities of the net.

LCD Liquid crystal display, a common display technology in which a film can be electrically changed from transparent to opaque. Characters are displayed by energizing the proper segments to form, for example, numbers and letters representing date and time on the face of an electronic wristwatch.

learn A neural net term denoting the process by which the net is enabled to solve problems. The net is taught a relationship, and thus it learns the relationship on the basis of data exhibiting that relationship. After learning the relationship, the net can then be consulted to solve new problems. Learning a relationship can be time-consuming, with intervals ranging from minutes to days, depending upon the computer and the size and complexity of the problem. Consulting the net, in contrast, is very rapid.

light-emitting diode (LED) A common indicator light technology providing higher efficiency, lower heat output, and longer life than the alternative miniature light bulb. Common LED colors are yellow, red, and green.

linear A characteristic of a mathematical relationship. This simple relationship may be expressed by the equation $y = Ax + B$. For given values of A and B, the relationship between x and y can be plotted as a straight line. If x were raised to a higher power, for example, or if there were a term in the equation involving both x and y, the relationship would not be linear. The computational complexity introduced by a nonlinear relationship is such that nonlinear relationships may be assumed to be linear, at least over a limited range, to streamline problem solving (at the expense of errors).

linear dependency A relationship between two parameters that is linear in nature. If one parameter is changed by X percent and another linearly dependent parameter changes by Y percent, then if the first parameter is changed by 2X percent, the second parameter will change by 2Y percent, and so on.

LISP A programming language developed for use in knowledge-based systems, with strengths in the manipulation of symbolic knowledge. Capabilities in rapid prototyping and knowledge manipulation are offset by system support challenges, due to the scarcity of LISP programmers. A counterpart to LISP, more popular in Japan and Europe, is another AI programming language called Prolog.

load data A functionality of the CAD/Chem product design tool, enabling data to be acquired from external databases or files.

manufacturing execution system (MES) An emerging software products area. MES represents the conceptual extension of MRP (**material requirements planning**) and **MRP II (manufacturing resources planning)** (*see also*), going beyond planning the materials and facilities requirements to manage the impact of production plans on the entire organization. MES calls upon plantwide computer networks and helps users balance the use of resources to best meet organization objectives.

manufacturing resources planning (MRP II) The conceptual successor to MRP (**material requirements planning**). MRP identified the raw materials and components required to support an anticipated production schedule. MRP II adds to this capability the ability to verify that manufacturing equipment and facilities are also available to support the anticipated schedule.

material requirements planning (MRP) A software tool that supports production scheduling and management. MRP provides the user a bill of materials for the materials and components that will be required to support the proposed production schedule. MRP is thus a planning tool.

membership function A method used in fuzzy logic to define control objectives. Rather than requiring that a parameter either is or isn't within a precise, or crisp, set, fuzzy logic enables a user to provide a degree of membership in a function. This degree of membership then determines the degree to which a control action is taken. The point of this approach is to accommodate errors in data and to enable easier statement of control objectives.

MES *See* **manufacturing execution system.**

model A mathematical representation of the behavior of a system. A model may be an intuitive understanding of a cause-and-effect relationship. But more commonly it is expressed mathematically and represents the result of analysis to determine how a system will respond to a range of inputs. A model might, for example, relate changes in process settings to changes in product quality, process energy usage, and so on. A model may be based on first principles, statistical methods, and now neural networks.

modular A design concept applicable to software and hardware. A modular design is one in which the complete system is composed of a number of modules. Each module has a specific function independent of other modules. Modular design provides system maintenance benefits because a single module can be revised or replaced without the risk of unintended consequences in other modules or functional areas.

MRP *See* **material requirements planning.**

MRP II *See* **manufacturing resources planning.**

multivariate regression Regression analysis to explore the interaction between multiple variables. Regression analysis is a statistical method that originated with studies of regressive (and dominant) traits in plants. Traditional regression analysis explores the impact on one variable of changes in the value of another. Multivariate regression analysis explores the impact on one variable of changes in the values of more than one variable.

net *See* **neural net.**

network 1. In the context of computer technology, a linking of computers and peripheral devices such as terminals, printers, and plotters by wire, fiber optic, infrared light beam, or radio. The objective is to enable users of computers and printers to share usage of hardware and share access to data and software tools. Networks may interconnect a small workgroup, or they may be international in scope and enabled by satellite communications. Common terms are LAN (local area network) and WAN (wide area network). 2. *See* **neural net.**

neural net A knowledge-based technology also known as neural networks and artificial neural network (ANN). Neural nets are a pattern-recognition technique. They can self-improve their performance and can discover relationships in data. This provides useful knowledge acquisition and modeling capabilities. Neural net-based tools have moved this new technology into design applications such as multiple objective design optimization and prediction and resolution of processing problems. Manufacturing applications include process optimization and quality improvements.

neuron In neural nets, as in mammalian brains, an element that functions as a summing junction and transmitter. Also known as a *node*. A neural net might consist of, say, 5 to 50 neurons, each interconnected to several others according to the particular neural net architecture employed. Each neuron adds together the signals presented as input and transmits an output signal on the basis of an internal transfer function, or relationship between input and output.

node *See* **neuron.**

noise 1. In general, that portion of a signal that is not desired because it is extraneous and interferes with the desired signal. When listening to the stereo, noise is the low-level hiss that is added to the pure music and is particularly noticeable as the only sound during quiet passages.

2. In Taguchi method terminology, noise is the effect of uncontrollable process deviations. The meaning is similar to the general meaning in that these deviations are uncontrollable and interfere with the goal of unvarying processes.

nondisclosure agreement Also known as a secrecy agreement or a confidentiality agreement. This type of an agreement is commonly formed between two organizations in a buyer–seller, partnership, or some other contractual relationship. Both parties agree that information from one or both organizations is valuable and therefore confidential and is to be treated carefully. This care typically includes nondisclosure outside of the organization and limited disclosure within. These agreements typically include a time limit of 3 to 10 years and describe the circumstances such as prior public disclosure or disclosure by others under which information is not considered confidential.

nonlinear A functional relationship that is not **linear** (*see also*). A nonlinear relationship between two variables, when plotted, is a curved line, unlike a linear relationship, which is a straight line. Many relationships are inherently nonlinear but, for ease of computation, may be assumed to be linear, at least over small ranges. If a linear relationship can be expressed as $y = Ax + B$, a nonlinear relationship might include x raised to a higher power as well as far more complex forms.

NO$_x$ Oxides of nitrogen, a pollutant that arises from combustion processes. The nomenclature indicates that it is a compound of nitrogen (N) and oxygen (O), but the number of oxygen atoms per nitrogen atom is unknown.

numeric Information expressed as a number. The alternative is symbolic information, which is information expressed as symbols. Words are a common form of symbol.

OEM *See* **original equipment manufacturer.**

open system A computer technology concept. The specifications of an open system of software or hardware are open, or available, to all system developers. This enables any developer to design and build a comparable product. The alternative is a proprietary system, in which only the developer of the proprietary system has access to the specifications of that system in sufficient detail to replicate the system. Proprietary systems thus limit competition and raise prices. These higher prices also make it easier for vendors to supply better product support than would be possible with an open system.

original equipment manufacturer (OEM) The manufacturer of the product bought by the ultimate consumer. The manufacturers of home electronics and automobiles are OEMs. The OEMs are the customers of the manufacturers who provide the electronic and mechanical components that make up the home electronics and automotive products.

optimize The act of finding the best possible solution, not merely an acceptable solution, to a problem. A number of optimization techniques exist, but their common goal is to enable a designer, for example, to find the one best solution and to know that the found solution is the best. The alternative is to try as many solutions as time permits and select the one that best meets the objectives.

output layer An element of a neural net architecture. The neurons in the output layer of a neural net provide the output, that is, the classification or estimation requested. One output neuron is used for each parameter for which an estimation is required. The neural net also incorporates an **input layer** (*see also*) and may incorporate one or more **hidden layers** (*see also*).

pattern recognition A problem-solving approach in which the situation is viewed as a collection of many pieces of information arranged in a pattern rather than as a collection of many unrelated pieces of information. Visual images represent a common pattern. The pattern recognition approach, when feasible, tends to be faster and more likely to recognize significant similarities while overlooking insignificant differences.

payback period In project justification, the period of time over which the sum of the reduced costs or enhanced revenues generated by the project will equal the cost of the project. All things being equal, a shorter payback period makes a project a more attractive investment.

PC *See* **personal computer.**

personal computer A small computer for home, office, or factory use. Although Apple Macintosh is a widely used personal computer, the term is often assumed to mean IBM or IBM-clone machines that use the DOS operating system and its successors OS/2 by IBM and Windows and NT by Microsoft.

piecewise linearization The representation of a nonlinear relationship by a series of linear relationships. When plotted, the piecewise linearization scheme approximates a curved nonlinear relationship with a series of end-to-end line segments.

PLC *See* **programmable logic controller.**

postprocess A term that refers to any processing of knowledge or data after some other operation, such as the generation of that data by a particular software package. Postprocessing can accomplish analysis, transformation, or presentation of data or knowledge.

preprocess A term that refers to any processing of data or knowledge before it is entered into a particular software package. Preprocessing can accomplish validation, analysis, or transformation of input data or knowledge.

programmable logic controller (PLC) A machine or process-control hardware scheme common in discrete part manufacturing and gaining acceptance in batch and continuous process applications. The basic PLC is an electronic replacement for relay logic, and as such it contains a program that analyzes inputs according to rules of logic now captured in software. This software replicates the wiring scheme used to implement logic with relays and uses a similar representation/programming language called *ladder logic*. PLCs are generally implemented in a more centralized manner than **DCSs** (*see also*) and incorporate I/O (input/output) modules to communicate with remotely located sensors and devices.

Prolog An artificial intelligence programming language especially popular in Japan and Europe, with LISP being more common in the United States.

qualitative From *Webster's Seventh New Collegiate Dictionary:* "of, relating to, or involving quality or kind."

quantitative From *Webster's Seventh New Collegiate Dictionary:* "1. of, relating to, or expressible in terms of quantity; 2. of, relating to, or involving the measurement of quantity or amount."

rapid prototyping The use of computer-based hardware to create a prototype part directly from a CAD **(computer-aided design)** file of the desired part. The objective is rapid development of prototypes to provide a design reality test and to check features such as appearance and fit to mating parts.

reasoning The use of logic or reason to draw conclusions. Expert systems incorporate rules to enable reasoning about situations so that problems can be identified and solved.

recipe The proportions and processing of ingredients combined in products made by batch or continuous processes. The recipe may represent a significant design problem if variations in the recipe, the raw materials, or processing affect the resulting product properties.

regression A statistical method to explore the impact on one variable of changes in the value of another. Regression analysis originated with studies of regressive (and dominant) plant traits.

repair data A functionality of the CAD/Chem product design tool. Data preprocessing, or repair, is performed to provide the fastest training and most accurate neural net-based design model. The repair data functionality advises the user on ways to fill gaps in data records and identifies duplicate or conflicting records as well as random or linearly dependent variables.

return on investment (ROI) In project justification, the effective rate of return, or compound interest, that cost savings or revenue gains created by the project represent compared to the initial investment in the project. Also known as internal rate of return (IRR), this calculation yields a percentage for which larger is better.

reverse engineer The task of generating a design in the form of paper or CAD **(computer-aided design)**-based drawings in reverse order, by starting with a finished but undocumented part. Reverse engineering may be required, for example, to repair or manufacture copies of parts for which original designs are no longer available.

robotics Multiple degree of freedom machines that may be reprogrammed to perform different motions. Robots are successful in applications such as welding and painting that require repeatable but reprogrammable motions, particularly in unpleasant working environments. Robots may be used in conjunction with computer vision or knowledge-based systems, but otherwise incorporate no machine-based intelligence.

ROI *See* **return on investment.**

rules The basis on which expert systems diagnose problems, identify requirements, and determine appropriate analysis and actions. Fuzzy logic systems use fuzzy rules to define control objectives.

sandcasting A common metal casting method in which a part is formed by pouring molten metal into a mold formed of packed sand. Design of the mold requires expertise, and the level of expertise and the geometry of the part determine part quality and process problems.

second order equation An equation in which variables are raised to powers less than or equal to 2, for example, $Y = Ax^2 + Bx + C$.

self-discovery The ability of a neural net to automatically discover relationships in data. This self-discovery capability can be used as a knowledge acquisition technique and as a way to enable a system to self-improve and adapt to changing circumstances.

sensor A device that provides, typically, an electrical output that varies in response to a particular process parameter. A sensor might measure temperature, pressure, flow, acceleration, or any of a large number of parameters. The electrical output might be a current source that varies between 4 and 20 milliamperes, or a voltage or resistance that varies over some range depending on the measured level. Alternatively, the sensor may provide a coded output in a format such as BCD (binary-coded decimal).

setpoint The target value of a process parameter. The goal of a process controller is to adjust the controlled variables (CVs) of a process to compensate for changing uncontrolled variables so that the difference or error between the setpoint (SP) and process variable (PV) is as close as possible to zero. The challenge is to respond smoothly and predictably under a range of operating conditions.

shareware Software provided at no cost, other than perhaps a nominal registration or documentation fee. This software is commonly acquired over computer networks. The low price is offset by lack of support and, often, restricted functionality.

shell *See* **expert system shell.**

signal analysis The analysis of, typically, an electrical or acoustic signal to extract information. Also called signal processing, this task may involve determination of the source or the propagation of a signal. A sonar signal might be analyzed to detect and identify submarines. Analysis of an ultrasonic test signal would reveal the size and type of internal flaws in metal. Analysis of a spectroscopy signal would indicate the constituents of a compound.

SPC *See* **statistical process control.**

SQC *See* **statistical quality control.**

statistical process control (SPC) The use of statistical methods to monitor and control processes. Measurements of process variables are taken on a regular basis. From these measurements, simple statistical measures such as the average value and range of these process variables are calculated. Operators or maintenance staff observe trends in these calculated values and adjust process setpoints to return the measured parameter values to the desired averages and to minimum range (or variation) of the values.

statistical quality control (SQC) The use of statistical methods to monitor and improve product quality. Measurements of product attributes such as dimensions or properties are taken on a regular basis. From these measurements, simple statistical measures such as the average value and range of these quality parameters are calculated. Operators or maintenance staff observe trends in these calculated values and adjust process setpoints, repair equipment, or identify changes in material characteristics as required to return the measured parameter values to the desired averages and to minimize range (or variation) of the values.

statistics The analysis of data to draw conclusions about the behavior of a system. Statistical methods are used to track and analyze current results of a system, using specialized methods such as statistical process control and statistical quality control (see also). Statistics-based models are also used for prediction of, for example, product properties or changes in process parameters. Alternative modeling methods include first principles and neural nets.

stereolithography One of several rapid prototyping technologies involving the fabrication of a prototype part directly from a CAD **(computer-aided design)** file of that part. The stereolithography process uses a laser to solidify liquid plastic in the form of the desired parts.

STL file A computer file containing a special format CAD **(computer-aided design)** required to drive a rapid prototyping process. The file format meets the needs of the rapid prototyping process beyond those required of a general-purpose CAD file.

symbolic A type of information based on symbols such as the words of spoken and written languages. The alternative is numeric information. The meaning of numbers is unambiguous, and their manipulation is straightforward. In contrast, symbols are often ambiguous in meaning, and as a result their interpretation and manipulation is difficult.

Taguchi method A statistics-based method primarily used to identify and rank factors affecting product quality. The objectives of this method are to minimize the number of experiments required to make this determination and ultimately to identify the least expensive way to achieve required quality levels.

total quality management (TQM) A management perspective that recognizes the importance of quality and customer satisfaction and calls upon all members of an organization to identify and resolve quality problems.

TQM *See* **total quality management.**

train An operation performed upon a neural net to prepare it to solve problems. The neural net discovers relationships in data that can be used as a problem-solving model. For example, a relationship may be discovered between the design and the performance of a product. The discovery of this relationship is called **learning** (*see also*), and a net that has learned this relationship is said to be trained. A trained net is then able to solve problems using the **consult** functionality (*see also*).

uncontrolled parameters Process parameters that are not controllable, even though they may vary and may have a significant impact. Examples of uncontrolled parameters include ambient temperature and humidity. One process optimization issue is to best meet process objectives such as quality and cost in the presence of variations in uncontrollable process parameters.

UNIX A common computer operating system, particularly for workstations. The operating system manages the internal operations of the computer, handling input and output functions, memory management, and the execution of programs. Operating system issues, from a user's perspective, include user friendliness and software compatibility. The flexibility and capabilities of UNIX are benefits for software developers and may or may not be seen as benefits by users depending on their interest in flexibility at the expense of simplicity and ease of use.

ultraviolet (UV) Radiation that is beyond violet that is, of shorter wave length than the shortest visible light (violet) and of longer wavelength than X-rays.

UV *See* **ultraviolet**

validate In the context of developing a model, the task of determining whether the model accurately represents the relationship being modeled. The goal of a **model** (*see also*) is to capture a functional relationship accurately and completely, so that the performance of a system can be predicted. A model can be based on **first principles** (*see also*), **statistics** (*see also*), and **neural nets** (*see also*). With each of these methods, rigorous development of a model will help ensure accuracy. However, that accuracy cannot be assumed until the model is asked to predict values for which the true answer is known, either by experimentation or experience. The comparison of predicted to actual values is called validation.

vision *See* **computer vision.**

VOC *See* **volatile organic compounds.**

volatile organic compounds (VOC) A class of organic chemicals that are of interest because they contribute to air pollution. As a result, the use of VOCs is increasingly controlled. Examples include gasoline, solvents such as paint thinner, and dry cleaning fluid. The control of VOCs is driving product design tasks such as the development of solventless paints and other coatings.

VMS A proprietary computer operating system (and trademark) of Digital Equipment Corporation, used on their VAX line of workstations and minicomputers. The operating system manages the internal operations of the computer, handling input and output functions, memory management, and the execution of programs. VMS is particularly popular in manufacturing applications, with its main competition in this area being the many versions of **UNIX** (*see also*) for workstation-based applications and **DOS** (*see also*) for PC-based applications.

weight 1. In the context of **neural nets** (*see also*), the magnitude of the multiplication performed by each interconnection between neurons. The values of the many weights are determined during the **training process** (*see also*) to enable the net to capture the desired functional relationship. The neural net equations, with the appropriate weights incorporated, explicitly describe the captured functional relationship. Thus, the neural net model is not a black box, or unknown transformation, but instead is described by these equations. These equations also enable the **consult** function (*see also*) to classify or estimate. 2. In the context of multiple-objective optimization, the ranking of a given objective, as stated by a weight or priority of that objective. The ranking of objectives is reflected in the proposed solution, in that if all objectives cannot be met, the objectives assigned the highest weight are most nearly met.

wireframe A CAD **(computer-aided design)** representation of a 3-D part. The wireframe model consists of line segments representing edges and surfaces of the three-dimensional part, as if the part were fabricated of wire mesh. An alternative representation is a solid model, in which surfaces are seen as opaque surfaces much like the real part. A wireframe is a computationally simpler if less realistic representation that also enables visualization of features that are internal or would otherwise be blocked by an opaque model surface.

workstation A class of computers exceeding the capabilities of PCs **(personal computers)** (*see also*). Workstations are characterized by an emphasis on computational power and thus are popular for design and manufacturing applications such as CAD **(computer-aided design)** and others in which complex calculations are involved. General business applications, for example, transaction processing—involve intensive transfer of data but less intensive calculation and, for this reason, represent a different market than those to which workstations are targeted.

Index